THE GREATEST BOOK OF ALL TIME

And the only source of comparative interval time data in the world . . .

.002 of a second: For a balloon to pop.

103 seconds: A blue shark to swim a mile.

15 minutes: A stripper to burn 36 calories.

4 hours: The Titanic to sink after ramming an iceberg.

38 days: For passage on a slow boat to China (New York to Hong Kong by freighter).

130 days: For the American worker each year to work off his portion of federal, state, and local taxes.

1 year: To build a Steinway piano.

Light to travel 5 trillion miles.

600 years: Oxford to admit women to degree programs.

One billion years: Petroleum to form.

"ENTERTAINING (a trained elephant can turn around on a stool in five seconds) . . .
EXCITING (a spotted hyena advances twice as fast as a cold front) . . . SLIGHTLY UNSETTLING (it takes .01 second for toothache pain to reach a cat's brain. The federal government probably spent $12 million to do that research.)"

The Hartford Courant

DURATIONS

The Encyclopedia of
How Long Things Take

STUART A. SANDOW
with CHRISSIE BAMBER
and J. WILLIAM RIOUX

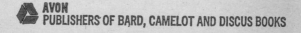

AVON
PUBLISHERS OF BARD, CAMELOT AND DISCUS BOOKS

AVON BOOKS
A division of
The Hearst Corporation
959 Eighth Avenue
New York, New York 10019

Copyright © 1977 by Stuart A. Sandow,
Chrissie Bamber, J. William Rioux
Published by arrangement with the author
Library of Congress Catalog Card Number: 77-7068
ISBN: 0-380-39859-1

First Avon Printing, September, 1978

AVON TRADEMARK REG. U.S. PAT. OFF. AND IN
OTHER COUNTRIES, MARCA REGISTRADA, HECHO EN
U.S.A.

Printed in the U.S.A.

ACKNOWLEDGMENTS

Most of the people we would like to acknowledge are all those named and unnamed data gatherers, some of whom are detailed in the resources section of this book. Others whose help was timeless and timely include Nancy Gross, who has always helped; Sharon Cuthbert, who contributed endless devotion to detail in the preparation of the manuscript; Karl Ege, for services above and beyond the call of friendship; Robert Weber, whose faith and interest in the project never lagged; Ann Buchwald and David Obst who smoothed the way and Robert Wyatt, our editor, who believed in the project and saw it through. Finally, we wish to thank all those loving members of our families who put up with the madness of it all.

—SAS, CB, and JWR

INTRODUCTION

We pursue time. It is a commodity so dear we buy it, sell it, borrow it, save it, and waste it daily in every way. But however finely we are able to measure it, and however well we use it, it continues to pass, unaltered as the bookends for our lives.

Durations has been compiled to let us know what continues while we spend our time. This is the first reference book that compares time data on the basis of common time intervals. Here we find no dates of wars or deaths of kings. These and epic events such as "revolution" and "renaissance" have been already well chronicled. Thanks to the advances of science, even the prehistoric evolution of the earth has found its way onto the time lines in history books.

Instead, we list here the infinitely repeated and repeatable. The facts that accompany the milestones of history but are seldom recorded as contributing to that history. We are speaking of the very pulse beat man carried with him as he emerged from the cave, the planting time he reckoned as he learned to grow things, the ten minutes in the morning reserved for shaving no matter what more memorable events would fill the day. Throughout these pages lies a panorama of things which fill time from the smallest to the largest increments found useful in our civilization.

Beginning with the time it takes light to travel

across a proton (.00000000000000000000001 of a second), we advance to the duration of one hundred successive generations of neutrons producing one billion, billion fissions (.000001 of a second), then to a lightning flash (.001 of a second), and on to the single stroke of the wings of a housefly (.003 of a second).

Steadily we proceed toward the one second mark and the events spanning our lifetime which we can clock ourselves without instruments more precise than our digital wristwatches. Finally, having leafed through the greater portion of the book, we turn the page on the upper limits of human time, and look at lifetimes of more than one hundred years, like the Giant Sequoia's; and then onto geological and astronomical data as unfathomable by the average person as the first entries. In this volume we end with the average life span of a planet (10 billion years).

Now there may be nothing remarkable about each of these entries in itself, at least nothing we have not encountered elsewhere, but as sets of facts occurring simultaneously, collected all in one place, their impact on us is different. For example, we never take the time to consider that while a dog bite victim waits out the three-day incubation for fulminating tetanus with fear, a dispassionate three-man crew is assembling a modular home. In the same twenty seconds it takes a cat burglar to make an illegal entry, hundreds of vacuum-packed wieners roll off a conveyer belt.

This *Encyclopedia of How Long Things Take* suggests the universe of possibilities as to how the commodity of time is spent. It can aid the speechwriter who needs to emphasize a point by timely comparison: "In the fourteen minutes I have been speaking, another forcible rape has occurred somewhere in this country." It is a resource for students needing time-related data on any of the more than 1000 subjects covered whether it be the ever-increasing speed of transportation in the last 100 years, man's attempts to reckon time in step with the solar system's, the biological rhythms of plants and animals, or the life expectancy of manufactured goods based on their planned obsolescence.

Perusing this volume, moreover, is a perfect way of "killing" time—while waiting in an airport, at the doctor's office or in the unemployment line. The charm and instructive qualities of *Durations* lie in the diversity of its entries whether the reader's intent be scholarly, practical, or mere trivia-gathering. We would be remiss, however, if we did not mention the several exceptions to this rule before moving on to explain briefly the reason we did this, why we need to know "how long things take."

While this volume has been an outgrowth of some scholarly work in the area of time orientation (or lack of it) of modern man, the text of the book is a straightforward presentation of time phenomena exclusive of any time theory. Yet, careful examination of this collection of time intervals can reveal a great deal about how man is unwittingly caught in the web of time.

In deciphering the book's message, let us first take note of the missing entries. They are, of course, infinite in number, for time is infinitely divisible and each division limited only by the nearly infinite number of occurrences which can be ongoing. If this seems a bit too cosmic for practical purposes, suffice it to say that the complete and unabridged version of *How Long Things Take* would fill the shelves of the Library of Congress or would require exclusive use of a computer bank. Obviously, in order to be approachable, this edition is highly selective. But this is not to say that it cannot be expanded. For the very point of presenting time as a series of experiences is that each of us has direct knowledge of a very limited number of time-related actions in which we are experts. There are readers who can fill these voids from their experience by telling us what they see, do, and create and the time it takes them. In this way there will be a host of new participants in subsequent editions of *Durations*.

Of the entries already cataloged, most fall within the human cognitive range of one second to one hundred years. Entries falling outside this range are almost uniformly technical or scientific. Most of these entries need microscopes, telescopes, or special photographic equipment to detect. This is a book compiled by humans about human time. If it had been written by gods who live forever there might be few entries of less than a million years, if by insects, no entries larger than six months.

Secondly, there are many more entries in the one to eight-hour section than the other periods of time that make up a day. We have been unionized to an eight-hour day and study or accept few tasks that cannot be accomplished within that time frame. Next, around each temporal milestone (one second, one hour, twenty-four hours, one year) there tend to be clusters of time-reckoning items, chronicling how man has attempted to commit to order once and for all the movement of the heavens and seasons throughout his lifetime. Consequently, the solar and sidereal days differ slightly from twenty-four hours, making us wonder when high noon really is and how Gary Cooper knew when to draw. The disparity between the Julian calendar and Gregorian calendar was so great (both battled for precedence in the early days of our republic) that George Washington at the age of twenty changed his birthday from February 11 to February 22, 1732, when Virginia adopted the Gregorian system in order to preserve his chronological age.

As the years progress, the entries change at forty years. Man is preoccupied with the illnesses he is apt to contract and how long it will take for them to wipe him out. Even before the forty-year mark, the underdeveloped nations of the world lead the procession to dusty death, with the earliest entry for the North American continent being its native Indian at forty-four years, his entire life expectancy now only one year longer than the working life expectancy for Americans overall. For most Americans it all ends in the eighties with the projected life span for clean-living members of the vegetarian Seventh-Day Adventists. Even without the intrusion of disease, scientists examining American men say it's all over at one hundred, that is when, like the refrigerator compressor programmed to blow after 10 years, the human body just wears out, the obsolescence date preplanned by our Maker.

"Time like an ever flowing stream bears all its sons away . . ." When the words to this familiar hymn were new, man had a clear perception of what a lifetime meant for him. He was to live longer than his beasts of burden but not as long as the oak seedling he planted in his youth. He might be laid to rest there beneath its branches, as sheltered as he had been in life by the scripture which ordered his days. Then in the 1700s, as our nation was in its embryonic stages, the time continuum was shattered as a result of an explosion in scientific thought. Man's sense of his own past and future had been secure when the Bible reigned as the ultimate historical authority on man and his planet. Now it had to be reassessed in the light of overwhelming scientific evidence. In the space of a few years the scope of history expanded from a scriptural six thousand years to a geological four and a half billion years. Amazingly, it took almost two hundred years after this disorienting experience to name the effect on our psyche "future shock."

The same burst of scientific thought which dethroned the Bible as historical authority, spawned the industrialization of western Europe and the New World. No longer did man's days flow in endless transition but, like the succession of locks along a man-made canal which change the water level, man came to think of his life as the sum of a number of clearly distinct parts. That legacy to us has been a dearth of language to describe adequately the nuances of gradual change but rather leaves us feeling that we exist in "chunks of now."

We are legally children until somewhere between seventeen and twenty-one then suddenly become adults to fit the legal needs of the state; middle-aged life begins at forty; then at sixty-five we become senior citizens because that is when the government phases us out of the work force and allows us to enter the movies at a reduced rate. Otherwise, society labels us worker, provider, pupil or welfare recipient and we are expected to act out our roles appropriately. But it is difficult to relate to mere labels. We fail to effectively communicate with each other on this basis. We become frustrated and alienated.

"Doin' time" is not a cynical view of our present

predicament. Don't we have a shorter workweek and more recreational opportunities than ever before? How about all the labor-saving devices the home-maker has? Why then do studies show that the work-week for the housewife has increased from forty-seven hours, six minutes in 1920, to forty-nine hours, eighteen minutes in the 1970s? Transportation of youngsters now accounts for an average of three hours, twelve minutes a week in a four-child household. That's something grandma didn't have to contend with and the breadwinner is moonlighting to pay for that second car and the lessons Mom drives the kids to. If he's not working two jobs then he is most likely going to night school.

The indicators are clear. We have gotten machines to perform much of the age-old drudgery, but we now use that "found" time for equally unpleasant tasks. Ironically, we have become extensions of the time-effective machines we build, operate, repair, and buy on time. The factory worker might even enjoy the same kind of work on his own time but in assembly-line production, even his breaks are geared to the convenience of the machines. This irony extends to control of interval time in general. Industrialized man thought he had harnessed time, instead it harnessed him and makes him trot through an endless succession of tiring paces. So, it is not unusual to hear a middle-class American bemoan the fact that he did not see or do as much on his vacation as he had planned. Under such a regime, "free time" fades into extinction and even leisure becomes harried. Time, once the great healer, has become the great disease.

This leads us to ask, what is this demon time and how has it eluded us? Time is the fourth dimension which colors our perception of the other three spatial dimensions of our world. Never has time so clearly defined space for us as in the current era of outer space exploration. The Moon, once nebulous and inapproachable, is now a round trip of seven days and provides a yardstick for measurement of the solar system in far more meaningful terms than sheer miles. If we can approach the speed of light, astronauts will bend time in their flight, returning after only a few years to find that the Earth has turned through eons of slower-paced solar time. Like latter-day Rip Van Winkles they will arrive home to find their loved ones, perhaps even their civilization, gone, only to ask the question, "What time is it, here?"

If time is truly relative as Einstein thought, then is it futile to search for a real sense of time? There undoubtedly will be special therapy needed for any of us caught in a time warp just as there are minor adjustments needed for jet lag, but for most of us who are still grounded most of the time, the dichotomy is less great, though just as crucial. If we can reconcile the inclinations of our biological time clocks with the demands of the clocks and calendars imposed by our society, we can reduce our anxiety about who and where we are.

There is only one alternative. If we cannot find our place in time, we can try to escape its boundaries. World-rejecting mystics like the present-day gurus have stopped or altered their perception of time through meditation. Likewise, drugs can either slow or accelerate time perception. Studies done with schizophrenics show that as they retreat into mental illness, they lose their sense of "present" in the past-present-future time continuum.

While some have sought comfort in withdrawing

from the constraints of time, they have concomitantly withdrawn from the real world. Conversely, to confirm the existence of the world as it is, is to explore it in the context of its relevant time concepts. It is the hope of those who present this book that we may better understand our lifetime by exploring time and through understanding, make better use of the time we have.

S.A.S.

DURATIONS

.000000000000000000000001 OF A SECOND
(10^{-24} SECONDS)

Light to travel across a proton

Unstable particle within an atomic nucleus to be created and destroyed

Life span of hadrons—*subatomic particles which hold the nucleus together*

.000000000001 OF A SECOND
(ONE TRILLIONTH OF A SECOND)

Light to travel .03 of a centimeter

A picosecond

.0000000004 OF A SECOND
(.4 BILLIONTH OF A SECOND)

The subatomic particle k-meson to form, travel, and disintegrate within an atomic nucleus

.000000001 OF A SECOND
(ONE BILLIONTH OF A SECOND)

Light to travel 30 centimeters

A nanosecond

.00000001 OF A SECOND

Fluorescence to take place in highly pigmented organic molecules

.0000001 OF A SECOND

Minimum time for light to impress a photographic plate

.000000157s
Minimum time in which the Barr and Stroud C.P. 5 continuous record camera can record a single frame

.000001 OF A SECOND

A microsecond

The maximum speed at which the Sultanoff camera can record a single frame—*fast enough to reveal the details of an explosion*

One hundred successive generations of neutrons to occur producing one billion, billion fissions

.000002s
To either put in or take out information from the main store of a computer composed of a ferrite core

.00001 OF A SECOND

In atomic explosions the interval between the emission of a neutron in one fission event and its capture by another nucleus

.000024s
A stick of dynamite to detonate—*pentolite*

.0000625s
A TV sweep circuit to produce horizontal light on a screen

.001 OF A SECOND

A millisecond

A main discharge of lightning to travel from the ground up to a thunder cloud

Phosphorescence to occur—*depending on substance. The afterimages on a TV picture tube are an example of observable phosphorescence*

.002 OF A SECOND

The first official telegraph message to travel from Washington to Baltimore

To take a flash picture with a German Zeiss lens set on its fastest shutter speed—*Zeiss lenses have their own shutter system*

Leicaflex SL to take a picture when set on its fastest shutter speed

The additional time each year that it takes the moon to circle the Earth

A balloon to pop

.0027s
A tuning fork on a watch to vibrate once

.003 OF A SECOND

A computer to lose its data in the event of power failure if data is not stored in a central processor

A housefly's wings to beat one stroke

.0044 OF A SECOND

The fastest shutter speed on the Kodak Pocket Instamatic camera

.005 OF A SECOND

A honeybee's wings to beat one stroke

.01 OF A SECOND

The sensation of toothache pain to reach the brain of a cat

A frog's muscle to begin twitching after electric shock has been administered

.02 OF A SECOND

A nerve impulse to travel from the toe to the head of a 6-foot-tall man

The minimum time before impact in an auto crash that the Nissan Motor Company's ESV will automatically tighten already fastened seat belts—*This Experimental safety vehicle has a pulse doppler radar sensor that measures the distance between the car and another object. If it's close enough for impact to be unavoidable, belts will tighten 20 to 100 milliseconds before the crash*

The amount of darkness following light that is needed to complete the cycle of photosynthesis

.03 OF A SECOND

An electric motor-driven microchronometer to indicate an interval of time—*accurate to 1/2,000 of a minute*

.033s
An electric beam to make 525 strokes of light across a TV screen

.04 OF A SECOND

The period during which a frog's muscle contracts after being shocked

An automotive air bag to inflate after impact

.05 OF A SECOND

The duration of response of human muscle to a stimulus such as a single electric shock (see also .1s)

A frog's muscle to return to normal length after contracting due to electric shock

.06 OF A SECOND

A modern knitting machine to knit a single course

Most electric motor-driven cameras to take a picture —*1,000 a minute*

.1 OF A SECOND

Minimum time for an echo to reach the human ear after hearing the original sound

An Olympic track star to cover 1 meter in a sprint

.15 OF A SECOND

The Apollo VIII rocket to cover each mile of its trajectory toward the Moon

.185s
To strike a typewriter key

.2 OF A SECOND

The joining time of plastics with ultrasonic welding

.206s
The space capsule to cover a mile in space while Ed
White was taking his space walk

.25s
Duration of a sixteenth note if the metronome is set
for 1/60

.28s
A tennis pro like Pancho Gonzales to serve—*balls
reach speeds of 112.88 mph*

.3 OF A SECOND

The first practical programmed computer introduced
in 1944 to add or subtract a 23-digit number

A steady-state reading to be reachable on a VU meter

.312s
To slam shut a drawer

.33–.5s
The balance wheel of a watch to oscillate

.36s
To pull open a drawer

To reach and grasp the top card from a deck

.4 OF A SECOND

To memorize each of seven nonsense syllables—*according to Hermann Ebbinghaus's classic experiment
on human memory*

.48s
Minimum time for the human eye to move from one
point to another and focus again

.5 OF A SECOND

To locate a record from an IBM magnetic disc storage
unit—*out of 50,000. The spindle revolves at a rate of
1,200 rpm*

Maximum reaction time of a major league batter to
a baseball pitch

An automotive air bag to deflate

The duration of an eighth note if the metronome is
set for 1/60

.552s
To turn a nut one revolution manually

.5935s
The male driver of an automobile to make the appro-
priate response to highway signals—*AAA study*

.6 OF A SECOND

To walk one pace

.668s
The female driver of an automobile to make the ap-
propriate response to highway signals—*AAA study*

.75 OF A SECOND

Duration of a dotted eighth note if the metronome is
set at 1/60

.8 OF A SECOND

The interval of prostrate contractions during the first
stage of human orgasm and during expulsion of
seminal fluid through the penile urethra—*William
Masters and Virginia Johnson study*

A jai alai ball to cross the court—*during play balls
reach speeds of 150 mph*

.82s
An adult to read a one-syllable word

.99 OF A SECOND

An adult to read a three-syllable word

HOW LONG DOES ONE SECOND TAKE?

1/86,400 of a day; since 1956 defined as 1/31,556,925.-9747 of the orbital year that began at noon on January 1, 1900

ONE SECOND

The shortest interval between windshield wiper strokes on 1975 Chevelle—*the controllable intermittent windshield wiper is optional on the new car*

A star to die and disappear—*minimum time*

A tuning fork pitched to A above middle C to vibrate forty times—*the pitch for musical instruments in the United States set by the National Bureau of Standards*

Eight million blood cells to die within one normal, healthy adult human

The wings of a small hummingbird to beat seventy times

Flames to spread 10 meters from the center of an explosion

An automobile traveling at 30 mph to go 44 feet

The primary shock waves from an earthquake to travel 5 miles

The Sun to travel 170 miles in its orbit and the Earth to travel 18.5 miles

A quartz crystal on a digital wristwatch to beat 32,768 times

Light to travel 186,282 miles through air and 124,000 miles through glass

A quarter note if the metronome is set for 1/60

The human eye to focus two to five times

Sound to travel at 20° C. 1,130 feet through air, 4,760 feet through water, and 16,400 feet through steel—*sound travels 2 feet per second faster per degree of centigrade*

A light bulb receiving alternating current at 60 Hz to go on and off sixty times giving the sensation of being constantly on

A .22-caliber rifle bullet to travel 1,200 feet

Telephone signals to travel 100,000 miles

The period a gymnast must hold a hand stand at the end of a stutz (turn) to be rated superior

A cheetah to sprint 34 yards—*clocked at 70 mph for sprinting only, it is the fastest mammal*

A race horse to cover 50 to 60 feet on a flat track

Material containing 1 curie of radioactivity to break down 37 billion atoms

To see twenty-four frames of film at the movies

Planets to travel in solar orbit (in miles): Mercury, 29.8; Venus, 21.8; Earth, 18.5; Mars, 15; Jupiter, 18.1; Saturn, 6; Uranus, 4.2; Neptune, 3.4; Pluto, 3

The average time between waiting for an answer to a question before the teacher moves to another student for the answer—*according to Ira Gordon, University of Florida, in a workshop at the National Council of Organization for Children and Youth Conference in Washington, D.C., February 3, 1976*

A meteor in space to travel 40,000 meters

The knifefish to emit 1,000 electric impulses—*used in the location of food*

The operating speeds of open reel tape recorders: 3¾ inches—medium high fidelity; 7½ inches—good high fidelity; 15 inches—professional quality

1,500,000 cubic feet of water to cascade over Khone Falls in Laos

A lightning leader to travel 290 miles toward the ground

1.188s
An adult to read a five-syllable word

1.2s
To make a field goal in American football—*including the snap from center, the ball placement, and an accurate hit by the kicker*

1.25s
Light to travel from the Moon to Earth

1.5s
The duration of a dotted quarter note if the metronome is set for 1/60

An 88.10-inch pendulum to swing—*computed for the latitude of New York*

1.5–2s
An earth boring machine to bore a hole one inch in diameter, three feet deep

1.8s
The interval at which an animal with electrodes hooked to the pleasure centers of his brain will press a lever in order to receive shock which seems to him pleasurable, often continuing at this rate of 2,000 times per hour for 24 hours—*B. F. Skinner study*

TWO SECONDS

Duration of a half note if the metronome is set for 1/60

To honor your partner before a square dance begins

To hear the echo of a sound which bounces off a wall 1,100 feet away on a cold day

Period of time a glass blower waits after heating a glass bulb red hot before attempting to blow into it

To be declared the winner in wrestling after pinning an opponent's shoulders on the mat

Norge, Wards, and Wizard brands automatic washing machines to stop spinning after the lid is lifted—*appliance cycle tests by Consumer Reports*

The acceleration of a free-falling object inside the Earth's atmosphere to reach a speed of 64 feet per second and to fall a distance of 64 feet

2.25s
To drive an 8-penny nail (3 licks)

2.2s
Period teachers will wait for a response to a question from boys (see also 7.7 seconds)—*study by R. C. Bradley, author of* The Role of the Principal in Driving Little Boys Sane

2.4s
The first stage of human male orgasm—*termed "ejaculatory inevitability" by William Masters and Virginia Johnson, it is the interval before emission when a man feels ejaculation coming and cannot control it*

2.5s
An echo to leave the surface and return from the deepest known part of the ocean (35,800 feet)

2.508s
An adult to verify a negative sentence as true or false —*experiment by Stanford University Professor Herbert H. Clark*

2.7s
The average duration of a cardinal's song—*from Arthur C. Bent studies in the wild*

2–3s
The duration of gray squirrel copulation

THREE SECONDS

To allemande left or do-si-do your partner during a square dance

To change a typewriter ribbon on a cartridge machine

The duration of a dotted half note if the metronome is set for 1/60

The acceleration of a free-falling object near the Earth to reach a speed of 96 feet per second and to fall a distance of 144 feet

To drive a 16-penny nail (4 licks)

3–6s
The fuses to burn on federally approved fireworks—*1974 regulations*

3–8s
The minimum time for a long distance call to go through when dialed direct anywhere in the continental United States—*variance dependent on an available switcher*

FOUR SECONDS

Goal time for saying the McDonald's Big Mac tongue twister: "two all beef patties, special sauce, lettuce, cheese, pickles, onions, on a sesame seed bun"

To develop film from the Longine camera that times photo finishes in sporting events

The acceleration of a free-falling object near the Earth to reach a speed of 128 feet per second and to fall a distance of 256 feet

The duration of a whole note if the metronome is set
for 1/60

4.055s
U.S. industry to consume a ton of alloy steel—
7,776,685 tons annually

4.8s
The "hang time" for a professionally punted football

FIVE SECONDS

Mice observed in the wild to copulate

The maximum time allotted for each move in the
fastest type of tournament chess

Safe speed for each person to pass through a revolving
door—*established by the National Fire Protection
Association*

The sound of thunder to travel 1 mile

A trained elephant to turn on a small stand

The period to dip garlic in boiling water so the skin
will slip off

To glaze eggs in a 350° oven when making *oeufs
miroir*

The acceleration of a free-falling object near the Earth
to reach a speed of 160 feet per second and to fall a
distance of 400 feet

The minimum time to turn over involuntarily in a
life jacket so that the victim's face will be out of the
water even if he is unconscious

5.4s
A mechanical picker to harvest a pound of cotton

5.8s
The first practical programmed computer introduced in 1944 to multiply a 23-digit number

SIX SECONDS

Most tents to catch fire after exposure to flame

Track star to run the 60-yard dash on an indoor track

The difference in speed records for men and women in competitive freestyle swimming for distances of 100 meters

A sky diver's static-line parachute to open

To absorb a pint of oxygen through the lungs while running

Cars built in 1900 to cover 100 yards (see also 30s)— *nearly a five-fold increase in speed over the first cars built only ten years earlier*

6.1s
To straighten and fold a terry towel (21"×17")

SEVEN SECONDS

Time lag for all "live" radio programs sent out over the airwaves in case on-the-spot editing is necessary

The period that Wolfgang Holtsmeyer keeps his head in a lion's mouth during his act with the Ringling Brothers Circus

A screen kiss during the romantic heyday of movies—
*The observation is made by Anita Loos, screenwriter
of the era of Mary Pickford, in her book,* Kiss Hollywood Good-bye. *This author is bemoaning the per-
missiveness of the '70s and longs for "one of those
tender, old-fashioned seven-second kisses exchanged
between two people of the opposite sex with all their
clothes on."*

7.4s
The half-life of the radioisotope of nitrogen

7.5s
The average adult to carry a 10-pound object 24 feet
and toss it aside

7.7s
Period teachers will wait for a response to a question
from girls (see also 2.2 seconds)—*study by R. C.
Bradley, author of* The Role of the Principal in Driving Little Boys Sane

EIGHT SECONDS

To qualify in a rodeo by staying on a bucking bronco
or bull in the riding events

To make a grand right and left around a four-couple
square dance set

8.5–10.5s
To be connected with the directory assistance operator
following the onset of a recorded message reminding
callers that the number may be listed in the telephone
directory—*Bell Telephone policy in some areas*

NINE SECONDS

Shortest Olympic performance on the balance beam (see also 2m)

TEN SECONDS

To give a walking-down cue on television—*seconds are counted off on the fingers to indicate to performers time remaining*

The period to dip white onions in boiling water to loosen skins

Bell Labs transaction telephone to make a credit card check for a merchant—*after inserting the magnetically coded card and typing the amount, the clerk waits while the phone automatically dials a credit agency's computer for verification*

To loosen the skin of a peach by dipping it in boiling water so that it can be peeled without using a knife

The period a professional basketball team has to advance out of its backcourt after the other team has scored

To present the shortest standard radio commercial—*air time*

To seize or kill anyone trying to steal uranium from a U.S. nuclear research reactor—*John Chancellor, NBC News Special, "The Nuclear Threat to You"*

A slinky to tumble down a full flight of stairs

The time allotment for making five shots in a rapid-fire pistol shooting competition

To put an animal "to sleep" with an overdose injection of anesthetic

To dehorn dairy cattle by applying direct heat from an electric dehorning machine

10.05s
An Olympic track star to run the 100 meter dash

TWELVE SECONDS

To land a commercial airliner—*landing 200 feet in the air, 3,500 feet from the end of a runway*

THIRTEEN SECONDS

To view one of Thomas Edison's kinetoscope "peep shows"

13.21s
For an Olympic track star to run the 110-meter hurdles

FOURTEEN SECONDS

14.7s
The first practical programmed computer introduced in 1944 to divide a 23-digit number

14–16s
To cut through an 8″×8″×16″ concrete block with an electric masonry saw

FIFTEEN SECONDS

The pasteurization of milk at 161° F.

Black and white Polaroid film to produce a print

The "Happiness Boys" to sing the interwoven sock jingle, the first singing commercial

The number of diagnosed cases of gonorrhea in the United States to increase by one

The number of burglaries committed in the United States to increase by one

Bees to communicate by dancing

To reduce pressure in a pressure cooker by placing it under cold running water

15–25s
The shortest change cycle time on a record changer (see also 25s)

15–30s
The germicidal action of chlorine to kill organisms

Minimum onset time of vaginal lubrication following sexual stimulation in women under fifty (see also 25s) —*William Masters and Virginia Johnson study*

SIXTEEN SECONDS

16.4s
U.S. industry to consume 1 ton of cotton—*1,920,480 tons annually*

SEVENTEEN SECONDS

A 1932 Duesenberg SJ to accelerate from zero to 100 mph

EIGHTEEN SECONDS

An electric masonry saw to cut through a 4-inch-diameter clay sewer pipe

The number of larcenies committed in the United States to increase by one

TWENTY SECONDS

A cloud to recharge after lightning flashes

Oil and vinegar to separate—*after thorough mixing at room temperature*

The difference between freestyle speed records held by men and women in competitive swimming for distances of 400 meters—*men being the faster group*

The maximum time an expert archer can hold a drawn bow without it quivering

A 1-inch bread cube to turn golden brown in oil that has been heated to 385°

To say 100 words—*shortest time in which most adults can articulate this number of words*

Monkeys who have had experience using sticks to figure out that they can reach bananas with them—*Koehler experiment*

To reduce a marshmallow to ashes by holding it directly over a red-hot flame

10–20s
Male chimpanzees to reach climax after initiating sexual relations

TWENTY-ONE SECONDS

A 1-cubic-yard dipper on a power excavator to pick up a load of dirt and deposit it in a dump truck

TWENTY-TWO SECONDS

Baseball players to perform any of the following recommended daily exercises using an Excrgenie: lats pulls, triceps pulls, pectorals, sit-ups, leg extension, and thigh pulls—*Exergenie equipment combines benefits of isometric and isotonic movement to maintain strength, endurance, and flexibility*

To memorize each of a list of thirty-six nonsense syllables—*Hermann Ebbinghaus's classic experiment on adult memory*

TWENTY-FOUR TO TWENTY-FIVE SECONDS

24s
The period of time an NBA team has to shoot after gaining possession of the ball

To sew one hundred stitches on the first sewing machine

25s
The period of time a college football team has to
start play after the ball is placed

TWENTY-FIVE SECONDS

The longest change cycle time on a record changer
(see also 15s)

25–85s
The shortest cycle of a home trash compactor (see
also 1m, 25s)

TWENTY-SEVEN SECONDS

A portion of spaghetti and meat sauce to heat in an
automatic vending machine introduced by the Kawasaki Steel Corporation of Tokyo

TWENTY-EIGHT SECONDS

A skilled operator to compute the following sum using
Sharp Electronics Calculator ⟶

```
    105.30  +
    216.00  +
    216.45  −
    105.60  +
    249.78  +
    289.00  +
    229.50  −
     38.64  +
    222.78  +
   1193.40  +
   1123.20  +
      9.01  +
    135.60  −
    594.60  +
    766.80  +
    266.40  −
    744.00  +
   1463.70  +
    558.00  +
    607.80  +
    879.90  +
     15.00  −
    778.20  +
    286.80  −
    333.60  +
     24.50  +
    222.25  +
   6621.42  +
  15997.73  *
```

THIRTY SECONDS

The metaphase of mitosis in the fruit fly embryo

An electronic fever thermometer to register an accurate reading (see also 3m and 5m)

Concrete to be shrink-mixed before the final mix enroute to the job

To wash a cow's teats

The first all-electronic computer, ENIAC, to accomplish the work which took a mechanical calculator 20 hours—*Electronic Numerical Integrator and Calculator*

One quart of water to seep through an uncoated, saturated cinder block

Whales and elephants to copulate—*not together, of course*

Diphtheria-causing bacteria to be killed in ice cream by exposing it to temperature of 65.5° C.

An individual hair follicle to be removed by electrolysis

The interval at which bottle-nosed dolphins come up for air while nursing

Maximum time a professional football team has to start play after the ball is placed

To create the following optical illusion: stare at an orange rectangle on a blue field before shifting the eyes to a blank wall where an image like the American flag appears—*The experiment is part of the exhibit in the Junior Museum in the Art Institute of Chicago*

Flying fish to complete a "flight"

Minimum spraying time daily with a vinyl chloride propelled bathroom spray that will cause cancer in rats—*according to the Environmental Protection Agency*

The usual duration of hypnotism in sheep

A model to apply lip color

15–30s
Maximum onset time of vaginal lubrication following sexual stimulation in women under fifty (see also 15s) —*William Masters and Virginia Johnson study*

The germicidal action of chlorine to kill organisms

30–60s
A couple to have a session of sexual intercourse marred by premature ejaculation—*defined by William Masters and Virginia Johnson as the time insufficient to satisfy the partners 50% of the time*

The local anesthetic tetracaine to reach its highest level in the blood after an intravenous injection

Minimum time for decay-causing acids to form in the mouth after eating sugar or carbohydrates (see also 1m, 30s)

THIRTY-ONE SECONDS

31.7s
Interval between fires in the United States—*1973*

THIRTY-FOUR SECONDS

The number of auto thefts committed in the United States to increase by one

THIRTY-FIVE SECONDS

A computer to make the 17½ trillion calculations necessary to come up with a personalized diet based on the dieter's favorite foods—*a service of The Favorite Foods Diet of College Park, Maryland*

To mass butcher a hog

THIRTY-SIX SECONDS

The interval at which rewarded animals will continue to press a lever to receive the reward—*B. F. Skinner study*

THIRTY-SEVEN SECONDS

To deal a deck of fifty-two cards—*according to the time and motion studies of G. B. Bailey*

FORTY SECONDS

A fire truck in the range of 25,000 to 30,000 pounds to accelerate from zero to 50 mph

For the brain to be affected by 10% after an intravenous injection of a barbiturate anesthetic

FORTY-THREE SECONDS

Number of violent crimes (murder, rape, robbery, or assault) committed in the United States to increase by one

FORTY-FIVE SECONDS

An A. B. Dick Magna I electronic typewriter to type a full-page letter

An Olympic track star to run the 400-meter hurdles

FORTY-EIGHT SECONDS

A helicopter to cruise 1 mile

An Olympic swimmer to swim 100 meters freestyle

FIFTY SECONDS

A fire truck in the range of 35,000 to 40,000 pounds to accelerate from zero to 50 mph

FIFTY-ONE SECONDS

51.42s
The swordfish or broadbill to swim a mile

FIFTY-TWO SECONDS

Minimum time to reach the 86th floor observatory of the Empire State Building by elevator (see also 1m, 25s)

FIFTY-SEVEN SECONDS

An Olympic swimmer to swim 100 meters backstroke

FIFTY-NINE SECONDS

Orville and Wilbur Wright's plane to cover 852 feet during the first powered flight in 1903

SIXTY SECONDS

30–60s
The local anesthetic tetracaine to reach its highest level in the blood after an intravenous injection

A couple to have a session of sexual intercourse marred by premature ejaculation—*defined by William Masters and Virginia Johnson as the time insufficient to satisfy the partners 50% of the time*

ONE MINUTE

A fire pumper discharging from a single 2½-inch hose line to deliver 1 ton of water

A skilled knitter to make 100 stitches by hand

An automated knitting machine to make 4 million stitches.

For the brain of a newborn baby to grow 1 to 2 milligrams

Cars on modern assembly lines to move 20 feet by means of conveyors and monorails

The baking time per ounce for small pike

Prompt initial radiation to be measurable following a nuclear explosion

Entire procedure of courtship and egg laying among sticklebacks (fish)

A draft horse to exert 33,000 foot-pounds of energy— *about 20 times that which a man could sustain (the standard for horsepower)*

A closing machine in a cannery to seal 600 to 1,000 cans

90 feet of 8½-inch-wide paper to be shredded by Destroyit Office Paper Shredder (model 6mm ¼″)

Allowable meditation time in public schools in Connecticut under a 1975 law

To cut 5 tons of coal from an underground seam using hydraulic equipment

The human heart to pump 12 gallons of blood during a period of physical exertion or stress

Milk letdown to reach its maximum after teat stimulation in dairy cattle

A human sperm to travel .1 of an inch

The human heart to pulsate 75 times

Iodine to kill bacteria (see also 15m)

For intermission between the first and second and the third and fourth periods in a high-school basketball game

The telophase of mitosis in the fruit fly embryo

The accuracy range of a quartz crystal watch during 1 year

The average adult to read 300 words

A pneumatic pile driver to sink a sheet piling for a basement foundation 2 feet through clay and 9 feet through sand

The rate at which people may safely leave a building during a fire drill or actual emergency: 60 people on a level; 45 people on a stairway or incline

The human male to produce 138,000 sperm cells

To get a color print developed with Polacolor film

One sandblast operator to cover the following materials: limestone, 7–9 square feet; marble, 3–5 square feet; granite, 6–8 square feet; terra cotta, 8–10 square feet; brick, 8–10 square feet; sandstone, 10–12 square feet

Average interval between swallowing in humans

The heart of a shrew to beat 1,000 times

To set 1,000 lines of type using an automated typewriter

The equivalent electric mixing time for every 150 strokes cake batter is beaten by hand

Minimum time for an air concentration of .64% carbon monoxide to cause headache and dizziness (see also 2m)

Minimum time for vaginal lubrication following the onset of sexual stimulation in women 50 years old and older (see also 5m)—*William Masters and Virginia Johnson study*

Hershey Foods Corporation to turn out 120 packages of pasta under the San Giorgio label (see also 1d)

An oarsman (single sculls) to make 40 strokes at the getaway

A ham radio operator to send a coded 13-word message —*standard of the FCC for obtaining an amateur's license*

To fill 340, 16-ounce bottles of Coke at the bottling plant

Minimum time for a residential water closet to refill after the toilet is flushed (see also 90s)

Minimum time for a human to stop bleeding after cutting a finger (see also 3m)—*3 mm.-deep wound*

1m, 5s
The elevator at the Washington Monument to reach the top observation level (about 500 ft.)

1m, 10s
The McCormick reaper to cut a 74-yard swath of wheat

Edward VIII to announce in a radio broadcast the abdication of the throne "for the woman he loved"

A white blood cell to attack and overcome a bacillus

Green Giant to flash freeze vegetables

The Bulova Photochart Camera to produce a developed picture of a photo finish

1m, 24s
A 1-cubic-yard capacity power excavator to load a dump truck with dirt—*truck capacity is four times the size of its dipper*

1m, 25s
Longest cycle of a home trash compactor (see also 25s)

Maximum time to reach the 86th floor observatory of the Empire State Building by elevator (see also 52s)

1m, 30s
The accuracy of a sand hour glass to vary 1 second

Each treatment for mental depression using electroconvulsive therapy (shock treatment)

A bluefin tuna to swim a mile

To heat a frozen dinner in a microwave oven (see also 20m)

Ultrafast silver fillings to set in the tooth to the extent that they can be carved

1m, 33s
To ride the roller coaster at Excelsior Amusement Park, Lake Minnetonka, Minneapolis–Saint Paul

1m, 35s
To go from downtown Seattle to Seattle Center by monorail

1m, 36s
The number of cases of diabetes recorded in the United States to increase by one

1m, 45s
An Olympic track star to run 800 meters

1.71m
A blue shark to swim a mile

1.91m

The median time for preadolescent males to reach sexual climax—*Alfred C. Kinsey, University of Indiana study*

TWO MINUTES

Baking time per ounce for the following fish: bluefish, croaker, herring, mullet, sea bass, or weakfish

The top burner on a gas range to heat a frying pan to 400° F.

To absorb a pint of oxygen through the lungs when the body is at rest

The period of radio contact with the crew of TWA Flight 514 on December 1, 1974 which shows their uncertainty about their low altitude before they crashed into a foothill near Dulles Airport (11:07–11:09 A.M.)

Amtrak's Metroliner to accelerate from zero to its cruising speed, 160 mph

The human heart to reach its peak rate after birth (174 beats)

A skilled hand seamstress to sew 50 stitches (see also 24s)

The interval during which three-fourths of the male population can reach climax—*Alfred C. Kinsey, University of Indiana study*

A dose of nitroglycerin to check an angina pectoris attack

To make shell molds at a foundry for bronze casting

To auction off a race horse

A dog show judge to find a fault on an individual entry
—*time limit at American Kennel Club shows*

Ducks to copulate

A giraffe to run a mile

Maximum time it takes the narcotic antagonist nalorphine to begin acting on narcotic depression (see also 1m)

Maximum time it takes an air concentration of .64% carbon monoxide to cause headache and dizziness (see also 1m)

The longer of two Olympic performances required on the balance beam (see also 9s)

2m, 1s
An Olympic swimmer to swim 200 meters backstroke

2m, 6s
The total eclipse of the sun on November 22, 1984 in Indonesia and South America

2m, 12s
The total eclipse of the sun on July 31, 1981 in Siberia

2m, 20s
Announcer Herb Morrison to describe as it happened the approach and burning of the *Hindenburg*

2m, 24s
Life span of a carbon steel blade cutting at a speed of 50 feet per minute

The population of California to increase by one—*an increase of 192,000 in 1973*

2m, 30s
U.S. industry to consume 1 ton of reclaimed rubber
—*210,092 tons annually*

The maximum duration of copulation between kangaroo rats

An electric range to heat a frying pan to 400° F.

Allotted time for an archer to shoot three arrows in official competition

Ten-year-old children in a testing situation to determine the true upright position in a tilted environment
—*research at the Down State Medical Center of the State University of New York*

To bake a potato in a microwave oven (see also 60–80m)

Maximum time to toast bread in most toaster ovens when the unit is preheated (see also 1m)

2m, 37s
The death toll due to cerebrovascular disease in the United States to increase by one—*200,000 annually*

2m, 42s
The total eclipse of the sun on February 26, 1979 in northwestern U.S. and Canada

2m, 45s
A harness racing horse to run a leisurely mile

2m, 48s
The total eclipse of the sun on October 12, 1977 in northern South America

2m, 54s
Interval between commercial interruptions on TV shows aimed at under-12 audience

Minimum time to launch 16 Polaris missiles (see also 4m)

Minimum time to complete a period in high-school wrestling (see also 3m)

Period a professional ice hockey team must play one man short if a misconduct penalty is called (see also 10m)—*A substitute can then be sent in*

To cook bacon in a microwave oven (see also 10m)

Minimum time of the arc resistance of melamine cellulose (see also 2m, 20s)

To kill diphtheria-causing bacteria in milk by exposing to temperatures of 55°–60° C.

Archibald MacLeish to read his poem "Epistle To Be Left in the Earth"

Minimum time to grind a single-vision lens for eyeglasses (per side) (see also 5m)

2m, 58s
To listen to the Ink Spots recording of "If I Didn't Care"

THREE MINUTES

A time-out in rugby—*which is declared only in case of injury during a game*

To steep tea

The V-2 rocket of World War II to travel to its target 180 miles away (its maximum range)

To do the Texas Star square dance

Glenda Jackson's divorce hearing

To cook a 3-minute egg after the water is boiling

Temperatures of 15,000° F. to penetrate a silicone shield like the ones used on space vehicles

To broil a one-half-inch-thick fish steak of the following varieties on each side: salmon, swordfish, tuna, halibut, or striped bass

To plump a raisin in hot water

To dissolve unflavored gelatin in warm water

To get an accurate temperature reading on a fever thermometer (see also 30s and 5m)

Minimum time for a quick-setting leak plug for masonry walls to harden (see also 5m)

Minimum time to succeed with the stimulation of energy with an acupuncture needle (see also 10m)

Minimum time of the courtship performance of the ruby-throated hummingbird (see also 4m)

Maximum time to complete a period in high-school wrestling (see also 2m)

Amtrak's Metroliner to come to a stop at a station after decelerating from a 160-mph cruising speed

To play an overtime period in water polo

To mount and balance a tire using modern equipment

The Russian jet dubbed "Foxbat" by NATO to fly 100 miles at 80,000 feet

One hundred twenty million tons of clay, soil, and gravel to slide into the sea after the 1964 earthquake at Anchorage, Alaska

A fire or rescue unit to put a vehicle on the road after receiving a call for assistance

A round in professional boxing

To anneal glass—*per millimeter of thickness at 67.5* microamperes

Beavers to copulate

An electric oven to preheat to 350° F.

A civil marriage ceremony

3m, 10s
An Olympic track team to run a 1-mile relay outdoors in good time

3m, 30s
The average "full service" fill up—12 gallons—check oil and water, wash windshield—*according to Tom "Jake" LaMotta*

To milk a cow producing 10 pounds of milk daily

3m, 55.9095s
To make up the difference between the sidereal day and the solar day

FOUR MINUTES

Each of six eating and watering periods a day on an automated chicken ranch

Maximum time of the courtship performance of the ruby-throated hummingbird

An opera singer to burn 10 calories while performing

A 3M copier to send a copy by phone

To pick a pound of cotton by hand (see also 5.4s)

Maximum time to launch 16 Polaris missiles (see also 2m)

A phantom jet to fly 100 miles

To boil vermicelli for use in casseroles

The prophase of mitosis in the fruit fly embryo

Minimum time for the human brain to die without oxygen (see also 7m)

4m, 18s
The total eclipse of the sun on February 16, 1980 in Africa and India

The minimum time to make a garment "permanent press" by curing it in an oven at the manufacturer (see also 18m)—*The time varies depending on fabric and finish*

4m, 30s
To milk a cow producing 15 pounds of milk daily

FIVE MINUTES

The maximum interval between contact with poison ivy and effective decontamination by scrubbing with strong soap and water—*After 5 minutes, resign yourself to an allergic reaction*

To castrate a cat (see also 30m)

To make water potable by boiling

To prepare minute rice

A hippopotamus to emerge from the water for air after submerging

Operating room personnel to scrub before any operations subsequent to the first of the day (see also 10m)

A cake to begin rising in a 350° oven

Average duration of contact with the cheapest prostitutes—*length of service for their basic rate*

To cook baked alaska

To kill the microorganisms responsible for botulism in canned pears by exposing them to temperatures of 115° C. (see also 30m)

The required addition of daily artificial light to the autumnal day (up to 6 hrs.) to increase the gonad growth in birds to springtime proportions

Period during which a woman must make four touches to win a fencing match (see also 6m)

To feel the somatic effects of LSD after an oral dose

A skylark to deliver a single sustained warble

Maximum time to grind a single-vision lens for eyeglasses (per side) (see also 2m)

The speaking time allotted each member of the U.S. House of Representatives in the Committee of the Whole when considering bill amendments

A marble factory to turn out 1,000 glass marbles

The human eye to adapt to a red light (see also 14m)

Human exposure to a concentration of 1:1,000 chlorine to prove fatal (see also 8–12h)

Intravenous novocain to take effect

Minimum time to die from a single 60 mg. dose of nicotine (see also 10m)

To be put into a hypnotic trance

A rest period in rugby

To activate dry yeast in warm (110°) water

To kill *clostridium botulinum* spores by exposing them to temperatures of 120° (see also 1h, 40m and 5h)

Maximum time to launch a Titan ICBM (see also 1m)

Minimum time it takes the North American Air Defence system (NORAD) to detect any Russian missile after launch (see also 15m)

Minimum time to begin feeling sick after drinking an alcoholic beverage if the alcoholic antagonist disulfiram is in the system (see also 10m)

Minimum time to soak glass, ceramic, or china coffee pots in a solution of chlorine bleach to remove stains (see also 10m)

5m, 24s
The total eclipse of the sun on June 11, 1983 in Indonesia

5m, 30s
Minimum time for most electric range ovens to heat to 400° F. (see also 6m, 30s and 9m)

5m, 36s
An American worker to earn enough to buy a quart of milk (see also 24m)

SIX MINUTES

The period during which a man must make five touches to win at fencing (see also 5m)

To cook a hard-boiled egg after the water boils

Minimum time to replace the rear wheel bearing on an automobile (see also 12m)

6m, 45s
To shoot or show 100 feet of Super-8 movie film

SEVEN MINUTES

To make a rush-hour tour of Lake Calumet in Chicago—*The boat, for those who cannot make the regular one-hour tours, runs from the Michigan Avenue Bridge to the Chicago and Northwestern Railroad Station*

Maximum time allowed to vote

Maximum time for the human brain to die without oxygen (see also 4m)

A king-size cigarette to burn without smoking it

7m, 30s
To play a chukker of polo—*six to eight to a game*

The death toll due to lung cancer to increase by one in the United States—*70,000 annually*

EIGHT MINUTES

To ride the wood-burning sternwheeler *Suwanee* around Greenfield Village at the Henry Ford Museum in Dearborn, Michigan

Light from the Sun to reach Earth

To complete a high-school wrestling match (three falls)

To cook live crabs in boiling water

To play a period of high-school basketball

8m, 54s
Human blood to coagulate at room temperature (two syringes: 2 ml. in each of two tubes, 8×75mm.)

NINE MINUTES

To drive through 4.3 miles of city traffic (see also 17m) —*study based on traffic in Hamden, Connecticut, after installing computer-activated control system. Reported in* The American City, *L. E. Davis, August, 1974*

The half-life of nonlabeled insulin in plasma (see also 40m)

9m, 30s
Maximum time allowable for commercials per hour
of prime time and on weekend children's TV shows
according to standards of the National Association
of Broadcasters—*down from 16 minutes per hour
before 1976*

The number of traffic fatalities due to collision in the
United States to increase by one

TEN MINUTES

To meditate on a deck of tarot cards to bring power
to the cards

A snowflake to form

To cook gefilte fish in a pressure cooker

Air conditioners to blow a cycle of fresh air throughout the Empire State Building

The interval at which new Model-A Fords rolled off
the assembly line using the new factory method for
mass production

Time between deaths of Americans due to cancer of
the colon

To knead a loaf of homemade bread

A typical period during which a duckling at critical
imprinting stage (16–32 hours, after hatching) follows
the mother or mother figure

Operating room personnel to scrub for the first operation of the day (see also 5m)

The rinse cycle on most automatic washers

The cones in the human eye to adapt to darkness (rods, however, take 30 minutes)

To declaw a pet

To cool bread or cakes before trying to remove them from the pan—*for the best results*

Methadone to be absorbed into the bloodstream

Duration of the first dream after falling asleep

A patch of fast-setting indoor cement to thicken

Minimum time to have a toe or finger amputated surgically (see also 20m)

To complete a daily dental regimen—*including proper brushing and flossing*

A misconduct penalty in ice hockey (see also 2m)

To beat a rompopé until cool (Nicaraguan rum drink)

To kill the diphtheria-causing bacteria by exposing it to temperatures of 58° C. (see also 1m)

To kill botulism toxins by boiling vegetables prior to home canning

Optimal time to bake, broil or poach fish one inch in thickness according to the Canadian method—*as reported by James Beard*

ELEVEN MINUTES

To make up the difference between the tropical year (from winter solstice to winter solstice) and the Julian year (365¼ days)

11m, 52s
Average wait for a city ambulance in the District of Columbia

TWELVE MINUTES

To make a simulated flight into space at the Spacarium at Seattle Center

Allowable time devoted to TV commercials during an hour of prime-time viewing on independent stations—*not affiliated with the National Association of Broadcasters*

Baking time per pound of cod fish in gas or electric ovens

Between breaths of the arapaima—*one of the world's largest freshwater fish*

A quarter of a professional basketball game

Minimum time to bake most cookies in gas or electric ovens (see also 15m)

Median duration of sexual intercourse—*according to 18- to 24-year-old males in a Morton M. Hunt study for Playboy Press*

A laborer to spread 1 cubic yard of loose sand or loam

To weave a yard of cotton cloth—*on a machine that can weave 180 picks per minute*

A bedbug to gorge himself on human blood

12.2m
An American worker to earn enough to buy a dozen eggs (see also 92.6m)

THIRTEEN MINUTES

Male song sparrows to cease singing after sunset (see also 14m)

13.90m
Number of deaths due to diabetes in the United States to increase by one—*37,800 annually*

FOURTEEN MINUTES

Number of forcible rapes in the United States to increase by one

The human eye to adapt to a white light (see also 5m)

To broil a 2-inch thick filet mignon rare (see also 17m)

FIFTEEN MINUTES

Extra time per pound to be allowed for roasting frozen meats

To open the mechanized concrete, steel, and lead door that protects the U.S. government's bombproof headquarters—*built inside a mountain near Berryville, Virginia*

A stripper to lose 36 calories during her act

To sterilize canning jars in boiling water

To boil lard and lye to form soap

Intervals between breaths of a hibernating jumping mouse

Interval after a seizure or fainting spell when the victim should not attempt to stand

A woman to complete the third stage of labor—*birth to expelling of the placenta*

To bring a gallon of water to a boil—*on most modern gas and electric ranges at high heat*

A period of an American football game

Minimum exposure time of a daguerreotype (see also 30m)

Minimum time to establish pumping and oxygenation outside the body with an artificial heart–lung machine before heart transplant surgery begins (see also 20m)

Minimum time for a surgical spinal anesthetic to take full effect (see also 30m)

Minimum time to die of cyanide poisoning

Minimum time for an antihistamine drug to relieve an allergy (see also 30m)

A scoop of ice cream to melt—*at room temperature*

To recover from the fumes of an aerosol protection device

For the flavor of milk to change by exposing it to daylight (oxidation)

For normal kidneys to excrete 25% of phenolsulfon- phthalein dye (see also 2h)—*used to test renal function*

Sound off interval for clocks with Westminster chimes

Period allowed U.S. congressmen to arrive for a roll call vote

Foreplay—*among single people under 25 years old; Morton M. Hunt study for Playboy Press*

To fuel and launch an Atlas missile

The safe period of sun exposure on the first day at the beach and the amount of time to be added each subsequent day

To knit a nylon stocking—*containing 925,000 stitches*

To cause hearing damage by exposure to sounds of 117 decibels (see also 30m and 90m)

Life of a neutron—*The time from the neutron's independence from the atom due to nuclear bombardment to the time the neutron disintegrates into a proton and electron*

To dig a grave with mechanized equipment

Period a tourniquet may be used to stop bleeding without damaging circulation

Minimum period for trying to revive someone with mouth-to-mouth resuscitation (see also 20m)

To soak silver clean—*in an aluminum pan with a solution of boiling water and baking soda*

SIXTEEN MINUTES

To cook a 5-pound roast in a microwave oven (see also 2h, 55m)

The most extreme variance of solar noon

16.67m
Light to travel across the orbit of the Earth—*186 million miles*

16.4m
An American worker to earn enough to buy a pound of butter (see also 140m)

16m, 40s
To burn 200 calories while running or walking up and down stairs

SEVENTEEN MINUTES

FBI trainees to run a measured 2-mile course over rough terrain

A Russian worker to earn enough to buy a loaf of bread (see also 5.4m)

EIGHTEEN MINUTES

Maximum time to make a garment "permanent press" by curing it in an oven at the manufacturer (see also 4m)—*time varies depending on fabric and finish*

To replace an automobile distributor cap

Maximum wash cycle on automatic washing machines

Foreign land–based missiles to explode on the east coast of the United States—*after the warning sirens are activated*

Commercial time during a telecast of a professional football game

Deep quiet period of sleep for the newborn

To boil lasagne noodles until tender

NINETEEN MINUTES

An American worker to earn enough during each 8-hour workday to pay for his recreational expenses

Zebra foal to attempt its first steps—*observed in the wild*

TWENTY MINUTES

Death by hanging

Foreplay—among single people 25–34 years old; *Morton M. Hunt for Playboy Press*

To freeze a gallon of homemade ice cream

To melt 200 pounds of bronze in a foundry furnace

A snake to consume a chicken egg four times the diameter of its own head

A Bessemer converter to produce 20 tons of steel

To kill any living organism at 121° C.

A woman who has previously given birth to complete the second stage of labor (stage of expulsion)

To pasteurize beer—*140–145° F.*

Optimum attention span of elementary school children—*according to John I. Goodlad, dean of the graduate school of education at the University of California, Los Angeles*

To steam a lobster

Minimum time for bacteria cells to divide by fission (see also 30m)

A half in a college basketball game

A human infant to nurse

Period during which burn victims should be bathed in cool water until they can be transported to a hospital

Two men to erect a stop sign

Minimum time for the benefits of methadone to begin —*after an oral dose*

A period of a professional ice hockey game—*three periods to a game*

To sterilize ear-piercing equipment—*to prevent the spread of viral hepatitis*

Maximum time for trying to revive someone with mouth-to-mouth resuscitation (see also 15m)—*then give up hope*

A carpenter to install a door stop

Minimum time for good paint remover to loosen the finish on furniture (see also 30m)—*so the old finish can be scraped off*

Recording period of human brain waves with an electroencephalograph

Half-time intermission of a professional American football game

To perform a hemorrhoidectomy

Minimum drying cycle of automatic dishwashers (see also 30m)

Hunger pangs to disappear after the first mouthful of food reaches the stomach

TWENTY-ONE MINUTES

Half-life of the alkali metal francium

Active REM-like sleep in newborns

TWENTY-THREE MINUTES

To hear one side of the first long-playing records— *introduced in 1948 by Columbia*

An American worker to earn enough during each 8-hour workday to pay for his medical expenses

TWENTY-FOUR MINUTES

A mechanic to adjust an automobile clutch pedal

Shortest comprehension time of a 60-minute tape recording adjusted by a VSC (variable speech control for tape recorder)—*This electronic marvel can compress or expand the normal 125–175 wpm speaking voice on tapes for people with special listening needs*

A Russian worker to earn enough to buy a quart of milk (see also 5.6m)

TWENTY-FIVE MINUTES

Staten Island ferry to make its trip from Battery Park to the borough of Richmond

Maximum time to have a lump excised from the breast (see also 15m)

A carpenter to install a latch hook, dead latch, door handle, and cylinder

A fashion model to set her hair on electric rollers and put on her face—*working at top speed*

To sear a roast—*This method of cooking begins by exposing the meat to high temperatures to seal in the juices*

A half in women's lacrosse

25m, 4s
To read the article "The Silent Drama Speaks"—*in the Spring 1975 issue of* Liberty *magazine. Most articles in the newly issued* Best of Liberty *from years past are time keyed*

TWENTY-SIX MINUTES

Daily reading of the newspapers by Americans who have not finished high school—*Alexander Szalai study*

Primary shock waves from an earthquake to pass through the earth

To see the silent movie *The Tramp*—*starring Charlie Chaplin*

TWENTY-SEVEN MINUTES

To make a flipper-dinger—*handcrafted toy (a pith ball is blown from atop a wooden pipe through a hoop). Time based on making toys in quantity by Dick Schnacke, handcrafter of mountain toys*

TWENTY-EIGHT MINUTES

A laborer to backfill 1 cubic yard of soil

THIRTY MINUTES

An expert to flay a hog by hand

A dog-sled team to cover 5 miles at top speed

Maximum time for bacteria cells to divide by fission (see also 20m)

To safely stop swelling of a bruised area with ice packs—*continued use results in a frostbite effect*

Maximum time for the benefits of methadone to begin (see also 20m)—*after an oral dose*

To spay a dog or cat

To "season" new pans—*first grease the pan, then place it in a 450° oven*

Maximum time a patent attorney is allowed to argue an inventor's case in a patent appeal

Irish toast: "may you be in heaven a half hour before the devil knows you're there"

Interior latex wall paint to dry

The human eye to adapt fully to a darkened room

A chicken to ovulate after laying

Nitroglycerin effects to wear off—*if taken to prevent angina attack prior to a period of stress or increased activity*

Five-mile ride on the world's longest route for a miniature train at the Fort Worth Texas Zoo

To reach a level of deep sleep

Minimum period of peak intensity for a haboob (desert dust storm) (see also 1h)

Daily contribution to household tasks made by children 6–11 years old—*New York State College of Human Ecology (Cornell U.) study 1973*

To play both sides of the 12-inch record introduced by RCA Victor in 1930—*the groove was 1½ miles long*

A stag to mount a doe twelve times—*observed in the wild, she ran away eleven times!*

Aspirin to be absorbed in the bloodstream

To pasteurize milk at 145° F. (see also 15s)

Side effects of a toxic oral dose of iron to manifest themselves after ingestion

Minimum time to feel the effects of marijuana resin after ingestion (see also 1h)—*more potent than smoking the leaves*

Sugar to be digested and absorbed into the bloodstream

An oral dose of ethchlorvynol to take effect—*this hypnotic drug causes feelings of depression*

Maximum time for a surgical spinal anesthetic to take full effect (see also 15m)

Maximum time for an antihistamine to take effect (see also 15m)—*for the relief of allergies*

Minimum telophase of mitosis in human cells (see also 60m)

Human coitus if the male is suffering from incompetent ejaculation—*William Masters and Virginia Johnson study*

THIRTY-ONE MINUTES

31.07m
Average duration of morning erections in humans 41–45 years old—*Alfred C. Kinsey, University of Indiana study*

THIRTY-THREE MINUTES

Daily reading of the newspapers by college-educated Americans—*Alexander Szalai study*

Number of annual murders in the United States to be increased by one

THIRTY-FIVE MINUTES

A speed reader to read an entire issue of *Time* or *Newsweek*—*claim of Evelyn Wood Reading Dynamics*

Average wait for a book to be delivered to a reader at the Library of Congress

To travel from Newark to Trenton by the Penn Central Railroad's 107 Metroliner—*a distance of 48.1 miles*

THIRTY-SIX MINUTES

To balance both front wheels on an automobile

Weekly yard care by housewives—*University of Washington study, 1970*

To make an Oscar Mayer wiener

THIRTY-SEVEN MINUTES

Louis Bleriot to cross the English Channel

FORTY MINUTES

Maximum time during which organs from cadavers can be transplanted after death of the donor (see also 30m)

A guided tour of the U.S. Capitol

An American worker to earn enough during each 8-hour workday to pay for his transportation—*Tax Foundation, Inc.*

The half-life of insulin I in plasma (see also 9m)

To play the 12-inch phonograph record developed by Thomas Edison in 1927—*its groove was 1½ miles long*

Daily transcendental meditation (two, 20-minute periods)

To shovel 1 cubic yard of sand by hand

The activity cycle of a 2-week-old baby—*as indicated by respiration, EEG's, body movement, and eye movement*

A guest at a cocktail party to consume a drink after the party has been in progress for 2 hours—*The Wine and Spirits Guide 1974*

To play a tennis match with sets at 6–1, 6–0, 6–0 (see also 3h, 30m)

40.62m
Duration of morning erection in humans 36–40 years old—*Alfred C. Kinsey, University of Indiana study*

FORTY-ONE MINUTES

Bulgarians who live in cities with a population of fifty to one hundred thousand to commute daily

FORTY-THREE MINUTES

Those whose family income exceeds $15,000 to travel to and wait to see a doctor—*(compare with 66 minutes and 81 minutes) Karen Davis, "A Decade of Policy Developments in Providing Health Care for Low-Income Families"*

43.9m
U.S. industry to consume 1 ton of natural rubber—*717,250 tons annually*

FORTY-FIVE MINUTES

Maximum length of most dreams

To boil cotton cloth during the dyeing process

Maximum time to repair torn cartilage in the knee joint surgically (see also 30m)

A guided tour of Harvard University—*the fastest way to get through college!*

To play a half in a soccer game—*no time-outs*

Installation of an apartment telephone

A ham to bake after glazing

45m, 30s
Daily care of pets by their owners—*Philip G. Hammer study, University of North Carolina, 1972*

To bake most casseroles in a gas or electric oven

Minimum time to surgically remove a gallbladder

Minimum time to bake an apple in a gas or electric oven (375°) (see 2m)

Minimum time to perform a hysterectomy (see also 1h, 30m)

FORTY-EIGHT MINUTES

Americans to travel to and from work daily—*Philip G. Hammer study, University of North Carolina, 1972*

To balance the wheels and tire assembly on the rear end of an automobile chassis

A professional basketball game

FIFTY MINUTES

Philippe Petit to make his 1,350-foot walk on a tight-rope between the twin towers of the World Trade Center—*1974*

The Moon's daily lag behind the Sun—*results in high tide occurring 50 minutes later each day*

Bread to rise until double in size (the second time) (see also 1h, 30m)

A woman giving birth for the first time to complete the second stage of labor (stage of expulsion)

Longest regular cycle of human fetal activity (see also 30m)

To get a suntan in June at high noon—*the point at which stimulation of pigment formation is at its peak*

FIFTY-THREE MINUTES

Daily personal care by American adults—*this includes bathing, grooming, and cosmetic application*

53.09m
Duration of morning erection in humans 26–30 years old—*Alfred C. Kinsey study, University of Indiana*

FIFTY-FIVE MINUTES

Two men to remove an 8-by-10-foot plasterboard wall

A session of psychoanalysis with Sigmund Freud

FIFTY-SIX MINUTES

A carpenter to program and drill 100 evenly spaced holes using a numerically controlled machine (see also 8h)

FIFTY-EIGHT MINUTES

An American worker to earn enough during each 8-hour workday to pay state and local taxes

FIFTY-NINE MINUTES

The time Americans allot to taking a walk (when they walk)—*Philip G. Hammer study, University of North Carolina, 1972*

ONE HOUR

Federal Reserve banks to sort sixty thousand checks—*"read" by high-speed electronic machine*

Daily listening to the radio by Americans—*Philip G. Hammer study, University of North Carolina, 1972*

To put through an intercountry telephone call in most parts of Europe

For 10 ml. of alcohol to be metabolized by humans

To prepare two hundred hors d'oeuvres—*one catering service employee*

An adult male to shed 600,000 particles of skin

Water to ascend 4 feet within a plant's stem

Daytime watch at the Tomb of the Unknown Soldier at Arlington National Cemetery (see also 2h)

Maximum time of the telophase of mitosis in human cells (see also 30m)

Maximum life span of an adult mayfly

A dehumidifier to remove a pint of water from a room —*at 70° F. and 70% humidity*

To de-ice 10 miles of a 2-lane road—*by spreading rock salt from a truck*

To screen print 2,000–3,000 consumer product packages for food and detergents using automated equipment

Maximum nightly homework given Russian first-graders

To take the College Board achievement test in any subject

To remove 400 pigs' hides automatically—*using a pig-skinning machine developed by Wolverine Worldwide (maker of Hush Puppies)*

Maximum time for the hunger-inhibiting factor of tobacco to wear off after smoking one cigarette—*The Pharmacological Basis of Therapeutics*

Maximum time a U.S. representative may speak in debate in the House of Representatives—*except by unanimous consent*

Daily contribution to household tasks by American teenagers—*Philip G. Hammer study, University of North Carolina, 1972*

Parker Brothers to print $392,000 in Monopoly money

Electronic switching systems of the telephone company to connect 100,000 calls

Tree sparrows to feed their young sixteen times

To regain full driving faculties after consumption of each ounce of alcohol

An active woodchuck to breathe 2,100 times—*while hibernating he breathes only 10 times per hour*

Winds to travel according to the Beaufort scale—1–7 miles, light wind; 8–12 miles, gentle wind; 13–18 miles, moderate wind; 19–24 miles, fresh wind; 25–38 miles, strong wind; 39–54 miles, gale wind; 55–73 miles, storm wind; 74 or more, hurricane

The Earth to turn 15 degrees on its axis

Period that Rome and Berlin are ahead of Greenwich Mean Time

To die from exposure in the waters of the Arctic Circle

For radiation to reach a peak of 2,000 roentgens following a nuclear bomb explosion—*20-mile radius of explosion*

A human sperm to reach an egg—*travel time*

Sound to travel 764 miles at sea level—*expressed as Mach 1*

A B-58 bomber to travel 1,528 miles (Mach 2)

A B-70 bomber to travel 2,292 miles (Mach 3)

A cold front to advance 20 miles

A warm front to advance 15 miles

A spotted hyena to run 40 miles

These volumes of paper to be shredded by the following models of Destroyit Office Shredders: 2-Way, 250 lbs.; Super Speed, 500 lbs.; Compact-Conveyer, 2,500 lbs.

To produce the following effects of alcohol consumption: ¾ oz.—hardly influenced, normal actions; 1 oz. —elation, sociable but incompetent driver; 2–3 oz.— lowered inhibitions, impulsive behavior; 5–6 oz.— confusion, staggering, slurred speech; 16 oz.—stuporous; 24 oz.—unconsciousness, death—*United Services Automobile Association*

The digitalis drug deslanoside to begin to produce its maximum effect after an oral dose (see also 2h)

ONE HOUR TO ONE HOUR TEN MINUTES

1h, 6m
Citizens classified as "poor" but not on welfare to travel to and wait to see a doctor—*(compare with 81 minutes and 43 minutes). Karen Davis, "A Decade of Policy Developments in Providing Health Care for Low-Income Families"*

Average time between eruptions of the geyser Old Faithful—*one of the world's most regular geysers, it rarely varies more than 21 minutes in its cycle*

1h, 7m
To prepare (12m) and bake (55m) orange-glazed lamb chops with mint-and-raisin pilaf

Daily time American parents spend in child-centered activities—*Philip G. Hammer study, University of North Carolina, 1972*

1h, 9m
Period that Americans will watch television at a single sitting (see also 92m and 111m)—*Alexander Szalai study*

ONE HOUR TEN MINUTES

To thaw a loaf of bread in a 200° oven

Two men to install a 10- to 12-cubic-foot, no-frost refrigerator (see also 1h, 25m)

To cook and cool Yugoslavian wine liqueur

ONE HOUR TEN MINUTES TO
ONE HOUR TWENTY MINUTES

1 h, 15m
The membrane in a chicken egg to be formed

To see the movie *I Was a Teenage Werewolf*

To prepare (15m) and bake (1h) pepperoni and chick-pea casserole

To prepare family meals for one day (see also 1h, 55m and 1h, 28m)—*done by women with college educations in a study of twelve countries including the United States by Alexander Szalai*

ONE HOUR TWENTY MINUTES

A paperhanger to hang wallpaper along a 10-foot wall (8-foot ceilings)

A roller derby—*8, 10-minute periods*

Two men to plant twenty privets in a hedgerow

ONE HOUR TWENTY MINUTES TO ONE HOUR THIRTY MINUTES

1h, 21m
Welfare recipients to travel to and wait to see a doctor —*(compare with 66 minutes and 43 minutes). Karen Davis, "A Decade of Policy Developments in Providing Health Care for Low-Income Families"*

1h, 21m, 40s
An average American worker to earn enough to buy a fifth of vodka

1h, 22m
Total period during which people dream during a night's sleep

1h, 24m
Primary or secondary conversation time of Czech men and housewives—*employed Czech women spend only 56 minutes. All the Czech figures are way below the world average (see also 4h, 24m or 2h, 6m). Study by Alexander Szalai*

Average weekly family taxi service by housewives— *University of Washington study, 1970*

The "period" of a pendulum the length of the radius of the Earth

1h, 25m
To visit a neighbor and not seem to rush in and out —*a study in Human Time Allocation at the University of North Carolina has determined the time that neighbors in suburban areas spend on any one visit*

1h, 26m
The half-life of barium

1h, 27m
The full preparation time of popovers

Stomach cancer deaths in women to increase by one in the United States—*6,000 annually*

1h, 28m
Preparation time of family meals for one day (see also 1h, 55m and 1h, 15m)—*by women who have a secondary education in a study of twelve countries including the United States by Alexander Szalai*

1h, 29m
The average length of time adults devote to reading a book at any one sitting—*Philip G. Hammer study, University of North Carolina, 1972*

ONE HOUR THIRTY MINUTES

The installation of a new residential phone and extension—*if underground wiring is completed*

To break a bucking bronco—*if the rider can stay in the saddle until the horse tires*

Hearing damage to be caused by continuous exposure to sounds of 103 decibels

The sleep cycle of an adult

To embalm a corpse

To crack and shell a pound of walnuts

To see the movie *How to Stuff a Wild Bikini*

Maximum amount of nightly homework given to Russian second-graders

To cremate a human body in a casket

Maximum observable time it takes Russian sables to copulate—*the longest period among mammals*

Bread to rise until double in size (the first time) (see also 50m)

Maximum parking time in Austrian cities in the "blue zones"

Minimum time to tune a piano—grand or spinet (see also 2h)—*they all have 88 keys*

ONE HOUR THIRTY MINUTES TO ONE HOUR FORTY MINUTES

1h, 32m, 36s
A Russian worker to earn enough to buy a dozen eggs

1h, 35m
Sputnik to circle the earth

1h, 36m
Time a camera must operate to read the many frames of a 4-minute animated cartoon

The average American worker to earn enough during each 8-hour workday just to pay his federal income tax—*Tax Foundation, Inc.*

ONE HOUR FORTY MINUTES

Amount of time devoted to church each week—*a study at the University of North Carolina has determined that Americans devote this amount of time to attendance of any one religious activity*

The *Tiros* weather satellites to orbit the earth

ONE HOUR FORTY MINUTES
TO ONE HOUR FIFTY MINUTES

1h, 42m
Two men to erect a fiber glass golf-tee shelter—*large enough to hold three men*

An American to earn enough to buy a man's shirt

1h, 45m
The difference between the longest and shortest days of the year at Hog Harbour on Espirito Santo, New Hebrides

1h, 46m
Time devoted by Frenchmen to eating each day—*highest world average for any nationality according to Alexander Szalai*

1h, 47m
Daily average that Americans spend doing nothing—*study at the University of North Carolina*

1h, 48m
To complete weekly sewing tasks in the average American family—*University of Washington study, 1970*

To fulfill a family social obligation—*Philip G. Hammer study, University of North Carolina, 1972*

1h, 49m
The viewing time of the Hitchcock thriller *Psycho*

ONE HOUR FIFTY MINUTES

To see the movie *The Secret Life of Walter Mitty*

Bobby Fischer to win nineteen games in chess in the Best of the World match in 1970—*five minutes was allotted for each complete game, Fischer winning 19 of 22 games played*

**ONE HOUR FIFTY MINUTES
TO TWO HOURS**

1h, 51m
A man in his twenties to burn 500 calories while engaged in light physical exercise

1h, 54m
The period devoted to a hobby on any one occasion—*Philip G. Hammer study, University of North Carolina, 1972*

The period that Americans spend watching any one sporting event—*Philip G. Hammer study, University of North Carolina, 1972*

1h, 55m
The average length of a meeting—*Philip G. Hammer study, University of North Carolina, 1972*

1h, 59m
To see the movie classic *Citizen Kane*

TWO HOURS

A person with a physical dependence on alcohol to experience withdrawal symptoms after taking his last drink

Period allowed to complete the undergraduate record exam

Newly hatched geese to begin following their mothers around

Two men to erect 100 linear feet of 4-foot-high snow fence

Ampicillin to reach its peak in the bloodstream—*an oral dose of the antibiotic*

Each of the twelve daily periods identified with animals of the Chinese zodiac on the Chinese calendar

A night watch at the Tomb of the Unknown Soldier in Arlington National Cemetery (see also 1h)

Symptoms of mushroom poisoning to become apparent after eating the poisonous variety *Amanita muscaria*

Dog watch on a ship—*4–6* P.M.

The best time to take penicillin after eating—*so food will not interfere with absorption*

The maximum a person may watch television daily and still be considered a "light viewer"—*criteria established by Dr. George Gerbner, dean of Annenberg*

School of Communications, University of Pennsylvania, leading authority on television violence

The autopsy performed on Howard Hughes

A 180° environment to sterilize potting soil—*such as the kitchen oven*

Aspirin to reach its peak effectiveness in the body

The amount cut from the U.S. workweek in each decade from 1850 to 1900

Bacteria to reach hazardous levels in perishable foods left at room temperature

Maximum amount of nightly homework given Russian third- and fourth-graders

To chill batter for making crêpe suzettes—*for best results*

To write a magazine advertisement estimated to yield $50,000–$100,000—*Joe Karbo of Sunset Beach, California estimated he would get this amount in $10 mail orders for his book* The Lazy Man's Way to Riches

The pituitary gland to respond to an injection of growth hormone

Television's family hour (7–9 P.M.)

Deslanoside to produce its maximum effect in the body (see also 1h)—*after an oral dose of the digitalis drug*

Two men to install an outdoor, wall-mounted telephone enclosure

Minimum time for the onset of symptoms of staphylococcus food poisoning after eating contaminated food
(see also 4h)

TWO TO THREE HOURS

2h, 10m
A track star to run a 26-mile, 385-yard marathon

2h, 12m
A parade float to travel the entire length of the Rose
Bowl parade route—*a rate of 2½ mph*

To polish "glamour colors" when painting automobiles

2h, 30m
Minimum time to perform a ureterocolostomy (see
also 3h)

To do a load of wash in 1917—*including operation
of manual washer, boiling, rinsing, and hanging time*

Length of a kindergarten day in New York State

To see a three-ring circus

To roast a 20-pound turkey in a microwave oven (see
also 7h, 30m–8h)

2h, 34m
An American worker to earn enough to pay his taxes

2h, 36m
To replace the cylinder heads on a six-cylinder car

For a man in his twenties to burn 500 calories

2h, 40m
Two plumbers to remove a fire hydrant (see also 6h)

2h, 42m
Englishmen to be helpful around the house on their
day off—*of all the nationalities participating in the
study at London University, the Englishman came off
as either the laziest or most efficient since he spends
less time than any of the others doing domestic chores*

2h, 55m
The biological half-life of LSD in man

THREE HOURS

A Duraflame log to burn

"Prime time" television—*defined as those programs
shown between 8 and 11 P.M.*

A group of clinically tested 7-year-old boys to experience a total of seven orgasms each—*Alfred C. Kinsey
study at the University of Indiana*

A laxative to work—*a saline cathartic like Epsom
salts should act about 3 hours after ingestion*

The accuracy of a spring-wound clock to vary 1 second

A Bowmar answer calculator to run down after a full
charge

The *Parade of Rex* at the New Orleans Mardi Gras

Alcohol levels in the blood and urine to reach their
peak after drinking an alcoholic beverage

Rest and activity cycle of the harvest mouse

The white of the chicken egg to be secreted around the yolk

The graduate record exam in any one subject

To complete a "watch" according to biblical reckoning

Daily free time enjoyed by U.S. women who live in rural areas—*Alexander Szalai study*

Maximum time it takes to die of cyanide poisoning (see also 15m)

Gymnasts to stay in top condition by daily practice five to six days a week

To pass through noontide according to the Saxon manner of reckoning days—*10:30* A.M.–*1:30* P.M.

The benefits of antihistamines to begin to wear off

Bon voyage party on an ocean liner—*the interval between the time when tourists may board ocean liners to inspect them and wish friends and relatives "bon voyage" and sailing time*

The boat ride around Manhattan

Peak alertness and work efficiency produced by 22 grams of protein

THREE HOURS TO THREE HOURS THIRTY MINUTES

3h, 2m
American men to complete their household chores on their day off—*a study by London University*

3h, 10m
To acquire a blistering sunburn in June at high noon
—*according to a U.S. government study*

3h, 12m
A mother of four children to provide "taxi service"
for her family each week—*University of Washington,
1970*

3h, 15m
To run out of gas in a Cessna Citation—*six-passenger
jet*

3h, 20m
A woman in her twenties to burn 500 calories while
walking, while a man of the same age can burn the
same number of calories by merely standing—*Food
and Nutrition Board of the National Academy of
Science and National Research Council*

An electrician to install an ultrasonic unit for a
burglar alarm system

3h, 23m
Americans who have sources of income other than
their main jobs to fulfill second-income obligations
daily—*Philip G. Hammer, University of North Carolina study, 1972*

3h, 24m
Daily period of free time that women with secondary
educations enjoy (see also 5h, 6m)—*Alexander Szalai
study*

3h, 25m
To ring every possible combination of eight cathedral
change bells

THREE HOURS THIRTY MINUTES

A set of dentures to "cure" during the process of fabrication

Two men to assemble a large backyard swing set

The laundering duties every week in a two-member household—*University of Washington study, 1970*

A tour of Universal Studios in Hollywood, including a 2-hour tram ride showing where and how movies are made and demonstrations of special effects

A seamstress to sew a woman's tennis outfit using the Stretch and Sew Method—*one of a series of advertisements stating time to sew Stretch and Sew garments*

THREE HOURS THIRTY MINUTES TO FOUR HOURS

3h, 40m
The viewing time of *Gone with the Wind*

3h, 48m
The average U.S. factory worker to log his weekly overtime—*U.S. Dept. of Labor Statistics*

3h, 59m, 51s
The recording time of excerpts from twenty-five of the Marx Brothers' radio programs on an album by the name "Three hours, fifty-nine minutes, fifty-one seconds with the Marx Brothers"

FOUR HOURS

The *Titanic* to sink after ramming an iceberg on April 15, 1912

The fragrance of perfume to begin to fade—*Mary Green of Parfums Givenchy recommends renewing fragrance at the pulse points of the body every 4 hours*

A ship's watch

An Easter egg roll on the White House lawn—*children accompanied by an adult are allowed on the mansion grounds for the annual event*

The offensive odor of a drowning victim to be eliminated once the corpse is placed in a freezer—*The American Way of Death (Jessica Mitford)*

A pacemaker to be recharged each month

Maximum time for the onset of symptoms of staphylococcus food poisoning after eating contaminated food (see also 2h)

A naturally ventilated house to cool off in the summertime if the evening temperature drops 10° F.—*time for inside temperature to equal outside (see also 2h, 20m)*

Time cut from the U.S. workweek in each decade between 1900 and 1940—*Although World War II brought increases in the workweek, they have been more than compensated for since its conclusion, and an additional 4 hours have been removed since 1945*

To butcher a side of beef manually

To unpack the belongings of the average family after a long-distance move—*done by a three-man moving crew*

A meringue pie to weep or wilt

The Earth to rotate 4 billion years ago—*theoretical*

Interval between infant feedings in U.S. hospitals

To soak brains before sautéeing

Maximum nightly homework given to Russian children in grades eight through ten

The antimalarial agent quinacrine to reach a concentration in the liver 2,000 times greater than the plasma level after administration

By definition, daily television viewing by a "heavy viewer" (see also 6h, 8m)—*criteria established by Dr. George Gerbner, dean of the Annenberg School of Communications, University of Pennsylvania, and nation's leading authority on TV violence*

To reverse the toxic effects of arsenic with the heavy metal antagonist dimercaprol

To melt 30,000 pounds of ingot metal in an oil-fired reverberatory furnace

A plumber to install a water hookup for an automatic washer

FOUR TO FIVE HOURS

4h, 6m
Housewives to plan and shop each week—*University of Washington study, 1970*

The quickest highly skilled auto shop technician can overhaul an automatic transmission assembly

4h, 12m
Average daily free time of Russians (see also 5h, 6m) —*Alexander Szalai study*

Two men to erect 100 linear feet of 3-foot-high cedar picket fence—*two rails*

4h, 30m
A daily session of the Supreme Court—*10 A.M.–2:30 P.M.*

To have a surgical face-lift

4h, 40m
To travel from Boston to New York by rail

4h, 48m
The lifeless star Cygnus X-3 to rotate around its partner star

4h, 54m
Time devoted weekly to dishwashing in households that have dishwashers—*University of Washington study, 1970*

Daily free time that men with college educations enjoy (see also 4h, 24m)—*Alexander Szalai study*

FIVE HOURS

Food to be completely digested in the human stomach

Length of public school day in New York State for grades one through six—*1975 ruling by Board of Regents*

A glass of carbonated soft drink to go flat

A cabinetmaker to make a drop-leaf end table

To soak the mineral stains off teapots by letting them stand in a solution of warm water and lemon rinds

To thaw each pound of a large turkey in the refrigerator (see also 50h)

To install a rolling bar-fronted door in a prison cellblock—*a four-man crew with a gas welding machine*

Cinderella's sisters to dress for the ball

Time limit for making forty moves in a chess tournament

Bulgarian men to work on chores around the house on their day off—*Of the nationalities studied at London University, the Bulgarians spend several hours more in household tasks than men in most other Western European countries or the United States*

Nuclear fallout to travel 75 miles downwind in the presence of a 15-mph wind

To make a loaf of Wonder Brand Bread

Minimum time for the U.S. House of Representatives to conduct its daily business (see also 6h)—*12 noon to 5 or 6* P.M.

Two men to paddle a canoe 20 miles in quiet water with no wind

FIVE TO SIX HOURS

5h, 3m
A woman in her twenties to burn 500 calories while standing

5h, 20m
Two carpenters to install a residential, prefab oak staircase (see also 20h)

5h, 30m
To bake an 8-pound ham

Length of public school day for children in grades 7–12 in New York State—*1975 ruling of Board of Regents*

The Coleman kerosene camp lantern to use a tank of fuel—*tank's capacity is 30 ounces*

A 3.5–4-hp outboard motor to empty its 2.5-gallon-capacity gas tank—*traveling at top speed with a light load*

Daily sleep requirement of a person 64–87 years old

5h, 33m
Period of daylight on the winter solstice on the Greenwich meridian at 60 degrees latitude

5h, 36m
A frog embryo to grow to 32 cells

5h, 48m
A paperhanger to wallpaper a 10-by-12-foot room

SIX HOURS

Maximum time for the allergy benefits of antihistamines (see also 3h)

Daily period during which the male ring dove incubates his mate's eggs

Daily practice of an amateur figure-skating champion

A Bessemer blast furnace to convert ore, coke, and limestone into molten iron

A packhorse or mule to travel 15 miles

A crew of three men to remove the stump of a mature tree from the ground

The Earth to turn 90 degrees on its axis

Maximum period of hospital confinement for the new "belly button" surgery technique—*developed at Johns Hopkins University Hospital for sterilization of women*

Kidney dialysis—*replaces natural kidney function, used three times a week for life*

Maximum period of work attempted by divers outside an underwater shelter—*but rebreathing apparatus is available for 10–12 hours*

Ampicillin to reach its peak in the bloodstream

Minimum time for a hurricane to form (see also 10d)

Average duration of barbiturate-induced sleep—*assuming the patient has not built up a tolerance*

Minimum time for symptoms of mushroom poisoning to become apparent after eating the poisonous variety *Amanita phalloides* (see also 15h)

Minimum time each day that employed Americans spend alone (see also 7h)—*Alexander Szalai study*

A bricklayer and helper to lay a 50-square-yard area with patio block

Decoration of a single Ukrainian Easter egg using centuries-old etching technique with stylus and wax

Period before flight time that domestic airline passengers are asked to confirm their return reservations

SIX TO SEVEN HOURS

6h, 8m
Average daily "turn-on time" for average American television set—*Nielsen ratings service, 1975*

6h, 15m
To roast a 22-pound turkey

To print a metropolitan daily newspaper—*based on four editions of* The Washington Post, *circulation 550,000*

6h, 24m
Weekly laundry chores for 3–5 people

Two men to install a six-foot-diameter whirling merry-go-round on a school playground

6h, 40m
To clean 1,000 used bricks with a pneumatic chipping hammer (see also 20h)

A skilled worker to install a dental light, floor- or ceiling-mounted

SEVEN HOURS

For a furniture factory to produce a bedroom suite

Average time to fabricate an artificial eye

To boil down apples to make apple butter—*demands constant stirring!*

A kidney transplant operation

The period that Denver, Colorado, and the rest of the Mountain time zone lag behind Greenwich Mean Time

SEVEN TO EIGHT HOURS

7h, 10m
Daily time Americans spend traveling on an out-of-town trip

7h, 15m
The Optimus kerosene lantern to use up a tank of fuel—*capacity is 37½ ounces*

7h, 18m
Weekly time American housewives spend cleaning house (see also 10h, 48m)—*University of Washington study, 1970*

Average interval between masturbations among males in early adolescence who have a record of high-frequency masturbation—*this age group is characterized by the lifetime peak for this behavior*

7h, 20m
To tear out 100 linear feet of railroad track and ties —*a four-man crew using earth-moving equipment*

7h, 30m
To fine grade, seed, and fertilize 1 acre—*four men and one piece of machinery*

To print a metropolitan Sunday newspaper—*based on four editions of* The Washington Post, *circulation 750,000*

To prepare an old-fashioned cheesecake—*including mixing ingredients (20 mins.), baking (1 hr., 10 mins.), standing in oven (2 hrs.), and setting period in refrigerator before eating (2 hrs.)*

7h, 34m
A woman in her twenties to burn 500 calories while sitting in a chair

A man in his twenties to burn 500 calories while sleeping

7h, 42m
The period of daylight on the winter solstice on the Greenwich meridian at 50 degrees latitude

7h, 46m
Americans to complete their workday—*excluding travel, according to a recent study at the University of North Carolina*

EIGHT HOURS

The period that San Francisco and Portland and the rest of the Pacific time zone lag behind Greenwich Mean Time

The exposure time of the first photograph—*taken in 1882 by Joseph Niepce*

One ram to successfully mount 114 ewes—*success rate determined by impregnation*

For the temperature of a newborn human to stabilize at 98°–99°

Two men to install an amusement park turnstile

A three-man crew to empty the average nine-room house of furniture prior to a long-distance move

One man to assemble 300 dresser drawers in a furniture factory

Weekly increased work load for housewives in homes where pets dwell—*University of Washington study, 1970*

Two carpenters to erect a 100- to 200-square-foot, free-standing, translucent swimming-pool enclosure

An adult to fulfill sleep requirements during a 24-hour period

Assemblage of a vest from Frostline kits—*an inside of goose down and an outside of ripstop nylon*

Repair time required by the telephone company's electronic switching systems during the next 40 years —*estimate*

A 285-ton "mole" tunneling machine to advance the Washington, D.C., Metro tunnel about 75 feet

Four men to lay the following quantities of sod in the eastern United States: 375 square yards on level ground, 350 square yards on slopes; in the Midwest: 1,900 square yards on level ground, 1,000 square yards on slopes

To have a permanent wave in 1906—*soon after it was developed*

Two structural steelworkers to install 280–300 square feet of window wall

A crew with a single air hammer to tear out: a 1,000-square-yard section of bituminous road or parking lot, 720-square-yard section of driveway, 330-square-yard section 6-inch, mesh-reinforced concrete, 20-square-yards of 6-inch, rod-reinforced concrete

A carpenter to lay 1-by-3-inch wood furring strips on the following materials: 475 linear feet of wood, 325 linear feet of masonry, 200 linear feet of concrete

One man to install a prefabricated fireplace

1 workday

To rough-in the average tract house after framing-in is complete

A plumber to install an autopsy table

A plantation worker to tap 600 rubber trees

To travel 40 miles in a horse-drawn carriage—*one day's travel*

NINE HOURS

Period that Tokyo, Japan, is ahead of Greenwich Mean Time

A tornado to run its course—*maximum life span*

NINE TO TEN HOURS

9h, 15m
The Coleman gasoline camp lantern, model 200A, to use a tank of fuel—*26-ounce capacity*

9h, 27m
Howard Hughes' flight from Los Angeles to Newark, N.J., in 1936 in his H-1 Racer, a record at the time

9h, 36m
To paint a 1975 Continental—*including preparation and polishing*

9h, 45m
To travel from Chicago to Memphis by rail—*Amtrak*

TEN HOURS

Maximum time to clip 200 sheep (see also 8h)—*an expert sheep shearer clipping the fleece in one piece (still quite a clip)*

A college student to read *The New York Times* cover to cover—*according to a* New York Times *ad*

Minimum period that food solutions stay in the large intestines

Half-life of penicillin G in the body with the additive anuria (see also 30m)

The sedative meprobamate to complete its period of effectiveness—*used to lower anxiety symptoms*

Two carpenters to assemble a revolving door

A horse-drawn plow to turn over 2 acres—*while a tractor-driven plow can turn over 20–40 acres in the same period*

To make a full set of false teeth

The number of hours per week that programs are broadcast over Bulgarian television

A child 5–9 years old to fulfill his daily sleep requirement

The installation of a 25,000-gallon septic tank

A Remington calculator to run down—*operates on four alkaline penlight cells*

The length of a complete area dog show—*duration of most American Kennel Club shows*

A camel caravan to cover 20–25 miles

The earth to turn 150 degrees on its axis

The Gutenberg printing press used by Ben Franklin to turn out 1,000, 4-page newspapers—*the original Gutenberg press could print 300 individual sheets during a 10-hour period*

To produce 28 tons of flour using 16 water wheels—*done in the Barbegal flour mills of the Roman Empire*

A combine to harvest the equivalent of 200–300 men working a 10-hour day

Minimum time to thresh 6–7 bushels of grain with a flail (see also 12h)

Minimum time to harvest 3–4 acres of grain by hand cutting it with a cradle (see also 12h)—*resembles a scythe with wooden prongs parallel to the blade*

TEN TO ELEVEN HOURS

10h, 4m
The period of daylight on the winter solstice on the Greenwich meridian at 30 degrees latitude

10h, 25m
To travel from Washington, D.C., to Atlanta by rail

10h, 35m
A crew and heavy equipment to install an 80-square-foot (minimum size) suspended highway sign over a thruway

10h, 36m
The daily period when women are alone—*time study by Alexander Szalai focused on women with secondary education in twelve countries—the period included sleeping time*

10h, 48m
The modern American housewife to complete weekly house cleaning—*University of Washington study, 1970*

Period of daylight on the winter solstice on the Greenwich meridian at 20 degrees latitude

ELEVEN HOURS

A carpenter to lay oak-strip flooring in a 12-by-12-foot room (see also 12h)

Period that New Caledonia is ahead of Greenwich Mean Time—*New Caledonia is an island in the Pacific Ocean*

Weekly household tasks American men perform—
*study at New York State College of Human Ecology
(Cornell U.) 1973*

A day on Uranus—*compared to an Earth day*

Daily sleep requirement of a child 3–5 years old

ELEVEN TO TWELVE HOURS

11h, 7m
Interval between masturbations among males at age
20 who have a record of high frequency masturbation

11h, 15m
To travel from Los Angeles to Houston by rail

11h, 18m
Period that women with incomplete primary educations spend alone (includes sleep)—*Alexander Szalai
study*

11h, 25m
The period of daylight on the winter solstice on the
Greenwich meridian at 10 degrees latitude

11h, 30m
Weekly time devoted to household tasks when children's play areas are poorly located—*so that the
mother cannot watch the children while doing housework, University of Washington study, 1970*

Average monthly time that volunteers work in U.S.
veterans hospitals

A crew of two carpenters and an electrician to assemble an 8-foot-square sauna

TWELVE HOURS

Daylight needed for poinsettia, aster, dahlia, cósmos, and chrysanthemum to bloom

To prepare for surgery by fasting

Maximum time to make steel from molten iron and scrap in an open-hearth furnace (see also 8h)—*100 tons at a time*

Onset of symptoms of *Clostridium perfringens* food poisoning after eating undercooked meat

The male indigo bunting to sing 4,320 songs

Discomfort from fasting to occur

Daily sleep requirement of a child 2–3 years old

The human body to reach its tolerance limit in water at 75° F.—*lowered body temperature. Blood sugar and muscle cramps signaled the tolerance limit in a group of nude men in the study by Blockley, West Virginia, 1964 NASA SP 3006: 128*

Seeds used for sprouting to absorb 60% of their mass in water by soaking

Maximum time to thresh 6–7 bushels of grain with a flail (see also 10h)

The Earth to turn 180 degrees on its axis

Rats to choose food over water after the onset of food and water deprivation (see also 72h)—*Douglas study*

A woman having her first child to complete the first stage of labor—*stage of dilation*

A caterer's staff of 280 to set tables for a dinner party for 4,000 people

Cream to rise to the top of fresh milk

A crew of three men to fell a 45–50-foot tree (20–24 inches in diameter), remove the stump from the ground, and cut all into 2-foot lengths

An LSD "trip"

Change cycle of weather preview flashed by the weather reporting signal of New York City's Mony Building

A deep coma due to acute alcoholic intoxication to endanger life

Minimum time to exhibit the acute gastroenteritis symptoms of salmonella food poisoning after ingestion of contaminated food (see also 48h)

The period of daylight on March 21 (vernal equinox) and September 23 (autumnal equinox) anywhere in the Northern or Southern Hemispheres

Period of sleep required to renew peak work efficiency after sleep deprivation of up to 48 hours

For 25% of an oral dose of the antibiotic ampicillin to be cleared by the kidneys

Pulped cherries to ferment

Minimum time for a morphine or heroin addict to fall into a restless sleep ("yen") following a dose of the drug (see also 14h)

Minimum time to recharge a dry cell battery (see also 16h)

12h, 55m
To travel from New York to Montreal by rail

THIRTEEN HOURS

Weekly food preparation for the average American family—*University of Washington study, 1970*

Weekly increased amount of time that mothers of children under the age of thirteen spend on household tasks—*as compared to mothers of older children*

Daily sleep requirement of human infants 6–23 months old

FOURTEEN HOURS

Yogurt making

Weekly time U.S. adults spend listening to radio music—*study by Broadcast Music, Inc.*

Daily sleep requirement of a human infant 3–5 months old

14h, 6m
Weekly food preparation in a family of 3–5 persons —*University of Washington study, 1970*

14h, 30m
"Working day" of the chipping sparrow—*during which he actively seeks food*

FIFTEEN HOURS

Interval between the evening and morning meals in many jails which serve only two meals during a 24-hour period

Maximum time for symptoms of mushroom poisoning to become apparent after eating the poisonous variety *Amanita phalloides* (see also 6h)

To assemble a radio direction finder for ships—*from a Heathkit*

Daily sleep requirement of an infant 1 to 3 months old

15h, 48m
Neptune to make one rotation on its axis

SIXTEEN HOURS

A skilled worker to install an electric or hydraulic dental chair

The ring kingfisher to incubate her eggs at one sitting

Factory workday for children—*before child labor laws*

Daily sleep requirements of a human infant during his first 15 days

Ducklings to reach the critical imprinting age

A day on Neptune—*compared to an Earth day*

To roof a house—*based on two men shingling a two-story home with basement totaling 24,000 square feet*

16h, 30m
A South African Airways 747 jet to fly from New York City to Johannesburg, S.A.—*SAA claims 2h, 30m faster than its competitor*

SEVENTEEN HOURS

The adult greater wax moth to die when exposed to temperatures of 46° C.

The larva of the yellow fever mosquito to die at temperatures of .5° C.

17h, 5m
To travel from New York to Chicago by rail—*via Washington, D.C., and Pittsburgh*

17h, 18m
Weekly meal preparation for a family of six—*University of Washington study, 1970*

EIGHTEEN HOURS

Minimum incubation of the common cold (see also 48h)

The satellite Janus to make one revolution around Saturn

Daily sleep requirement of the sloth

Marble factories to "cook" to the molten stage the silica, soda ash, and other ingredients that make up toy marbles

Uninterrupted period during which the female ring dove incubates her eggs

Effects of sleeping pills to completely wear off

Minimum period an organ such as the heart or kidney can remain viable outside its host (see also 24h)—*at 15° C.*

Workday of the American pioneer and homesteader

For the absorption process to be completed in a human large intestine

Minimum time for a large dose of a salicylate like aspirin to be 50% eliminated from the body (see also 2–4h)

NINETEEN HOURS

The Earth to turn 285 degrees on its axis

19h, 30m
The annual cerebral palsy telethon

TWENTY HOURS

To assemble a stereo cassette tape deck—*from a Heathkit*

Estimated time to crochet and fringe a prefabricated, four-panel afghan in a kit—*designed by Sammie Freund for Studio Knitting, Inc.*

A crew to install and equip a 26-foot-long, automotive spray-painting booth

Weekly gardening chores required by the floral clock at Edinburgh, Scotland—*The gardener weeds and maintains the more than 22,000 plants from a ladder covering the clock's face*

To clean 1,000 used bricks by hand (see also 6h, 40m)

Newly hatched ducklings to exhibit their first emotional response—*Fear*

To die of yellow oleander poisoning after ingesting its "nuts"—*most frequent cause of accidental poisoning in man in Oahu, Hawaii*

To travel from Los Angeles to New Orleans by rail

TWENTY-ONE HOURS

A fastidious family in 1917 to complete their spring housecleaning in a seven-room house: redressing maple floors—4 hours; woodwork and paper walls cleaned—3 hours; kitchen walls washed or painted—3 hours; bathroom walls washed or painted—2 hours; cleaning mattress and bedding—3 hours; cleaning furniture—3 hours; cleaning basement—3 hours—*Window washing was performed every month in the household from which this log was taken for a total of 4 hours. It might be noted that they lived near a foundry*

TWENTY-TWO HOURS

Minimum time to die from a lethal oral dose of iron tablets (see also 48h)

The Aurora Borealis or "northern glow" to be visible in Helsinki from May to July

22h, 30m
To read the New Testament aloud

22h, 50m
Three carpenters to install 14-by-6-foot truck scales

TWENTY-THREE HOURS

23h, 12m
An American to earn enough to buy a man's medium grade suit

23h, 40m
The satellite Mimas to make one revolution around Saturn

23h, 56m, 4.9s
A star to cross over the same meridian twice—*sidereal day*

HOW LONG DOES A DAY TAKE?

The time it takes Earth to make one rotation on its axis

TWENTY-FOUR HOURS

Maximum time for nerve gas G-agents to dissipate in the atmosphere (see also 10m)

Maximum time an organ such as the heart or kidney can remain viable outside its host (see also 18h)—*at 15° C.*

Effects of methyldopa on the heart rate and blood pressure to wear off after a single dose

Grasshoppers to eat 1½ times their weight in grass—*about .05 ounce*

To complete the circadian cycle

To die from an oral dose of arsenic

Mimosa leaves to open, close, and open again

Maximum period a juvenile can be held in detention or shelter care if no delinquency charge is filed

Trial period for giving medication for glaucoma

A baby blue whale up to the age of seven months to gain 200 pounds

A pair of house wrens to feed their young 1,117 times

To run Europe's leading motor race—*"the 24 hours of Le Mans"*

U.S. taxpayers to lose over $1 million in interest because the federal government keeps $3.9 billion of withheld taxes in non–interest paying bank accounts

An oral dose of isoniazid to be completely excreted through the body (see also 144h)—*Isoniazid is used in treatment of tuberculosis*

To have the oxygen tent removed following most types of surgery in the chest area if no complications develop

A mother whale to produce 200 quarts of milk

Ascaris females to lay 200,000 eggs—*species of the parasitic roundworm*

A cyclone to travel 200–500 miles in the summer and 700 miles in the winter

The average man (22–35 years old) to utilize 2,800 calories 65 mg. protein: 5,000 iu (international units) vitamin A; 30 iu vitamin E: 60 mg. ascorbic acid:* 18 mg. equiv. niacin: 1.1 mg. riboflavin: 1.4 mg. thiamin: 2 mg. vitamin B-6; 5 mg. vitamin B-12; .8 g. calcium; .8 g. phosphorus: 140 mg. iodine: 10 mg. iron; 350 mg. magnesium and undetermined quantities of vitamin D—*.4 mg. Folacin*

An adult to produce one to three pints of perspiration

The skin to begin peeling after eating quantities of polar bear liver—*The symptom is one due to Vitamin A poisoning which is possibly due to the abnormally high concentrations in polar bear liver*

Americans to write 95 million checks each business day of the year—*1971 figure of the Federal Reserve*

A 45-year-old man who weighs 154 pounds to utilize 2,600 calories and a 45-year-old woman who weighs 128 pounds to burn 1,850 calories

Seminal fluid volume to be restored to 3 ml. in young men following orgasm

Humans without liver function to die

The ten pints of blood in the human body to make 1,000 complete circuits

The average person to walk 19,000 steps—*8 miles*

Plaque to colonize on the teeth

Period every week when animals in the wild naturally fast—*they choose not to eat one day out of seven*

To get a divorce in Haiti or the Dominican Republic

The Pentagon to destroy ten to fourteen tons of "secrets" in its basement shredder

One hundred professional photographers to shoot sixty thousand pictures to be used in the LIFE magazine bicentennial publication, "One Day in the Life of America"

Hours per week that programs are broadcast over Yugoslavian television

To run up an operating cost of $36,000 at the New Orleans Superdome—*whether or not it is being used for a sporting event*

Mallee fowl, chicken-like ground birds, to begin to fly

The desalting plants of Kuwait to produce 100,000 cubic meters of freshwater from seawater—*This tiny country leads the world in freshwater production with an abundance of the petroleum products needed to run the plants*

One rooster to service his hens the following number of times: 2 hens, 24 times; 13 hens, 31 times; 26 hens, 46 times

Period of "irreversible coma" before being declared legally dead—*no electrical activity of the human brain*

The onset of salmonella food poisoning after eating contaminated food

An untreated corpse to begin to decompose

To withdraw from short-acting barbiturates and reach a stabilization level with pentobarbital

A supernova to release as much energy as the sun produces in a billion years—*supernova is a catastrophic stellar explosion*

The New York General Post Office to handle 20 million pieces of mail

A person to lose a pint of water by respiration—*exhaling it through his mouth and nose*

The accuracy of a tuning fork to vary one second

Marjorie Holmes to add another thousand words to her stock of religious and inspirational writings—
Holmes is the author of such best sellers as The Joys of Being a Woman *and* I've Got To Talk To Somebody, God

To shed 50–80 hairs from the human head

For baby chicks to begin feeding

One person to inhale 15,000 quarts of air

The human kidneys to filter and return to the bloodstream three times the entire body weight in water and salts—*about 200 quarts*

A queen honeybee to lay 1,500 eggs at the peak season of production

One cell of bacteria to produce 4,700,000 quadrillion offspring by fission—*theoretically, if there were enough food supply*

To help the average person better grasp the 4½-billion-year history of the earth, the *National Geographic Magazine* offered a comparison of the relative geologic time spans to a 24-hour day: 24 hours = 4½ billion years; 1 hour = 180 million years the ocean basins have been formed; 1 minute = 3 million years; 1 second = 50,000 years ago when cave men walked the Earth; .01 second = 500 years; .001 second = 50 years working lifetime of humans

To test skin sensitivity to hair dye with a patch test

24h, 19m
Strom Thurmond's filibuster of the civil rights bill—
August 28–29, 1957

24h, 37m
Mars to make one rotation on its axis

24h, 50m
The Moon to circle the Earth

24h, 52m
The cycle of tides

TWENTY-FIVE HOURS

Russian peasant women to make the traditional embroidered rubashka blouse

To burn enough calories to lose a pound of fat by walking

TWENTY-SIX HOURS

A cat intoxicated with phenobarbital to become hyperactive

A bricklayer and helper to build a masonry handball or squash court

TWENTY-EIGHT HOURS

Interval between masturbations among males at age fifty who have a record of high-frequency masturbation—*Alfred C. Kinsey, University of Indiana study*

TWENTY-NINE HOURS

Hours per week that programs are broadcast over Hungarian television

THIRTY HOURS

Ice to melt and the resulting water to reach 50° in a well-insulated cooler when the outside temperature is 90° F.

Average workweek reduction since 1850—*from 70 hours to 40 hours. This provides each employee with an extra 1,500 hours a year*

For an infant to exhibit left-to-right eye coordination —*study by John Watson*

Incubation of angel fish

30h, 42m
The estimated week of the average U.S. industrial worker in the year 2000

THIRTY-TWO HOURS

To assemble an FM tuner from a *Heathkit*

THIRTY-THREE HOURS

Two men to erect 100 linear feet of 16-foot security fencing around a prison

33h, 29m, 30s
Charles A. Lindbergh to fly solo from Mineola, New York, nonstop to Paris in 1927 in *the Spirit of St. Louis* (see also 6h, 38m)

THIRTY-FOUR HOURS

34.7h
Workweek of the U.S. worker in a wholesale or retail business

THIRTY-FIVE HOURS

35h, 50m
Workweek of U.S. workers in apparel and textile factories

THIRTY-SIX HOURS

To sell 120,000 tickets to a six-show series in New York featuring the rock group Led Zeppelin

Weekly household tasks in a home without children —*University of Washington study, 1970*

Continuous work without rest to result in 15% lowered work efficiency

To sprout sunflower seeds

Seminal fluid volume to be restored to 5 ml. in young men following orgasm

The anticoagulant benefits of sintrom (acenocoumarin)

Redness to appear as a positive reaction to a tuberculin skin test

Withdrawal symptoms to begin after a morphine or heroin addict is deprived of the drug

THIRTY-SEVEN HOURS

Hours per week that programs are broadcast over Soviet television

37h, 6m
U.S. worker to complete his workweek

THIRTY-EIGHT HOURS

38h, 30m
Hours per week that programs are broadcast over Polish television

THIRTY-NINE HOURS

Bacteria in pasteurized milk to double—*when stored in refrigerator at 40° F.*

39h, 54m
Workweek of U.S. workers engaged in the manufacture of furniture

FORTY HOURS

Mattress testing by "dynamic fatiguing"—*the process employed at the Good Housekeeping laboratory uses a 185-pound mannequin to compress the mattress five times a minute for a total of 12,000 times*

Playing life of guitar strings

To assemble a stereo receiver from a *Heathkit*

A fireman to complete his first aid training

Four carpenters to set up a lane in a bowling alley including the alley, pin setter, scorer, counter, and supplies

The average interval between incidence of coitus among married men of high-school age—*Alfred C. Kinsey, University of Indiana study*

40h, 18m
Workweek of U.S. workers involved with food processing and manufacture of electrical equipment

FORTY-ONE HOURS

41h, 42m
Average American workweek in 1950

FORTY-TWO HOURS

An average Soviet worker to earn enough to buy a woman's dress—*Sylvia Porter, 1973*

42h, 20m
Workweek of a U.S. miner

FORTY-THREE HOURS

Hours per week that programs are broadcast over Czech television

FORTY-FOUR HOURS

The workweek in Cuba

Continued work without rest to result in 20% lower work efficiency—*report of the American Industrial Hygienic Association*

FORTY-FIVE HOURS

The burning life of the first incandescent light bulb made by Thomas Edison

Hours per week that programs are broadcast over West German television

45h, 36m
Interval between incidents of intercourse among experienced married men 18–24 years old—*Morton Hunt study for Playboy*

45h, 42m
The total time devoted to weekly household tasks by women who work 26–39 hours outside the home

FORTY-SEVEN HOURS

47h, 6m
American housewife to complete all her household tasks in 1920—*U.S. Department of Agriculture*

47h, 36m
American housewife to complete her household tasks in 1952

FORTY-EIGHT HOURS

To see and do everything at Disneyland or Disneyworld

Fish and guests to turn sour—*according to an old saying*

Rosh Hashanah

Cellulose molecules to break down during the viscose process that turns wood pulp into synthetic fiber

Radiation following a nuclear explosion to be reduced from 200 to 20 roentgens—*the human body can absorb an accumulated dose of 100 roentgens and survive*

The ten-event Olympic decathlon—*five events are run each day*

The maximum storage period for fresh chicken and ground beef in the refrigerator (32° F.)

A Navajo medicine man to assemble the material used in a single ritual cure

Yeast to work in the brewing of ale

The waiting period between purchase and delivery of firearms in these states: Alabama, Indiana, Pennsylvania, and South Dakota

The cut life of roses without preservatives (see also 4d)

To forget 68% of a list of nonsense syllables to the extent that they can no longer be recited—*C. W. Luh study*

The annual herb, scarlet pimpernel (*Anagallis arvensis*), to produce death in sheep when fed at 2% of the animal's weight

Continued work without rest to result in 35% lower work efficiency

Total sensory deprivation under laboratory conditions to cause a decrease in color perception—*study by J. S. Vernon et al.*

To extract alcohol from carrots

An Atmos clock to run down given less than a two-degree temperature differential

Average hospital stay for varicose vein surgery

Recommended bed rest following tonsillectomy

Symptoms of belladonna poisoning to subside

Cattle to recover or die from inkweed poisoning (see also 18–24h)

Newborn calves to begin identifying and following their mothers

Maximum time for the incubation period of the common cold (see also 18h)

Maximum time to exhibit the acute gastroenteritis symptoms of salmonella food poisoning after ingestion of contaminated food (see also 12h)

Maximum time to die from a lethal oral dose of iron tablets (see also 22h)

TWO TO THREE DAYS

Young grouse to use up their food supply stored in unabsorbed yolk

To brew homemade ale

To cure a pneumonia patient with antibiotics

To sprout buckwheat, corn, or pumpkin seeds

The process of vinous fermentation to take place in the making of white wine

FIFTY HOURS

Sports participation during a 4-month period necessary to qualify for a presidential sports award in any one sport

To thaw a 10-pound turkey in the refrigerator

The playing life of a sapphire stylus on a turntable

FIFTY-ONE TO FIFTY-NINE HOURS

53h
The man-hours needed to assemble a prefabricated folbot—*English-style canoe*

57h
The average interval between incidence of coitus among married men who are 30 years old—*Alfred C. Kinsey, University of Indiana study*

SIXTY HOURS

The average tourist to see the sights in the nation's capital

The reading time for the average daily issue of the *Congressional Record—more than 4 million words*

The average workweek during the nineteenth century in industrialized societies

Est training (2–30 hour back-to-back weekend sessions)

Hours per week that programs are broadcast over French television

Interval between incidents of married coitus experienced by men 25 to 34 years old—*Morton Hunt study for Playboy*

60h, 28m, 48s
The satellite Ariel to make one revolution around Uranus

SIXTY-ONE TO SIXTY-NINE HOURS

64h, 12m, 36s
The half-life of the radioisotope gold 198

65h
Hours per week that programs are broadcast over East German television

65.8h
Mean interval between incidents of married coitus experienced by couples 25 to 34 years old—*Morton Hunt study for Playboy*

66 h
To fabricate and cure refractory cold-molded plastic capable of heat resistance up to 1,300° F.

68h
A skilled worker to pull 1 acre of flax by hand

Hours per week that programs are broadcast over Belgian television

The satellite Dione to make one revolution around Saturn

SEVENTY HOURS

The work week of a man with seven children, including full-time employment, household tasks, and child care—*study at New York State College of Human Ecology, Cornell University*

70h, 54m
The average weekly broadcasting time of U.S. educational television stations, 1973

SEVENTY-ONE HOURS

A body louse egg to receive lethal exposure to cold at −5° C.

71h, 2m
The longest Moon walk by earthlings—*U.S. astronauts Charles M. Duke, Jr., Thomas K. Mattingly and John W. Young in April 1972*

71h, 20m
To travel from New York to Los Angeles by rail—*with 9½-hour delay in Kansas City*

THREE DAYS

God to deliver the Ten Commandments to Moses on Mt. Sinai—*Exodus 20:11*

A bronze mold made of silica sand, cement, and water to set

Period during which an egg will lie on its side in a pan of cold water after being hatched (see also 10d)

Worker honeybee eggs to hatch into larvae

Thomas Edison to mold the first incandescent light bulb filament

The duration of German measles (rubella) rash

For the onset of fulminating tetanus symptoms after infection—*This form has nearly a 100% mortality rate (see also 4d and 7d)*

The miller's daughter to guess Rumpelstiltskin's name in the fairytale classic

A three-man crew to deliver, unload, and assemble a modular house on a foundation (three, 8-hour workdays)

A fertilized human ovum to grow to thirty-two cells

Effects of drugs used on dairy cattle to pass through milk—*milk should not be shipped before this time but discarded*

Seventy percent of the antidepressant imipramine hydrochloride to be eliminated from the system

Jesus Christ to rise from the dead—*Matthew: 27–28*

To air cure nonrefractory, cold-molded plastics which must be held to close dimensions

Scurvy-grass wine to ferment

Cut life of carnations without preservatives

To sprout chickpeas, peas, or sesame seeds

The stools of a newborn human to begin taking on the character of the food ingested—*before this time, meconium is excreted from fetal nutrients*

A Polish wedding party in the Tatra Mountains

The stagecoach ride from Manhattan to Philadelphia in colonial days

Maximum time the speaker of the House of Representatives can appoint a speaker pro tempore without a consent from the House

Enemies of St. Januarius to heat the furnace that was to consume him—*The saint refused to burn but emerged from the furnace feeling in the pink!*

Interval between the closing of entries and the horse race known as an "overnight race"

Period a consumer may rescind a contract (three business days, ending midnight on the third day) (see also 10d)—*Federal Reserve Truth in Lending Law gives consumer privilege of voiding a contract to buy real estate in which a security interest is or will be retained*

Jonah to get out of the belly of the whale—*Jonah 1:17*

For gravel to be present in young song birds after hatching

The period that a dog's kidney can be stored outside its host and then returned without loss of function

Average time between birth and onset of postnatal depression

Lamb chops to lose freshness in the refrigerator—*32°*

The stability of an unbuffered solution of penicillin to no longer be insured

A charge service such as American Express to receive charges from business establishments following purchases

Period before flight time that international airline passengers are asked to confirm their return reservations (if more than 72 hours between flights)

Waiting period between purchases and delivery of firearms in Miami Beach, Illinois, and Rhode Island —*no wait in the rest of Florida*

To make "distiller's beer"—*the fermented base from which whiskey is made*

Hansel and Gretel to find the witch's gingerbread house in the forest

Period following his birth when St. Rumbald of Kent preached continuously—*He then dropped dead*

To conche milk chocolate

To get a marriage license in Alaska, Arkansas, the District of Columbia, Florida, Indiana, Iowa, Kansas, Kentucky, Massachusetts, Michigan, Mississippi, Missouri, Pennsylvania, Tennessee, Washington, and West Virginia

Incubation period of influenza

3.1d
Average length of school suspension for Asian ethnic children in U.S. public schools—*Children's Defense Fund*

SEVENTY-THREE TO SEVENTY-NINE HOURS

75h
To ring every possible combination on a set of 10
cathedral change bells (see also 3h, 25m and 40y)

75h, 24m
The mean interval between incidents of married coitus
experience by men 35 to 44 years old—*Morton Hunt
study for Playboy*

76h
A crew with heavy equipment to build 100 linear feet
of running track on an athletic field—*gravel and
cinders over stone base*

EIGHTY TO NINETY-FIVE HOURS

80h
To weave a 6-by-4-foot tapestry on a hand-powered
shuttle loom

80h, 42m
To complete weekly household tasks in families who
have children under 12 years of age—*University of
Washington study, 1970*

84h
Hours that programs are broadcast over Peruvian
television weekly

The period of hospital confinement for having dilation and curettage performed

The mean interval between incidents of married
coitus experienced by couples 35 to 44 years old—
Morton Hunt study for Playboy

3.53d
Average length of school suspension for Spanish speaking children in U.S. public schools—*Children's Defense Fund*

92h, 10m
To read the complete Old and New Testaments aloud

94h, 22m
U.S.S.R.'s *Vostok III* to make sixty revolutions around the earth—*on August 11, 1962, to complete the first group space flight*

THREE TO FOUR DAYS

To ferment sage wine

The poliomyelitus virus to be detectable in the body after exposure

For danger to remain after the introduction of blister gas to an open area—*summer conditions*

Pain to subside following a mastoidectomy

THREE TO FIVE DAYS

Minimum time for beet seeds to germinate under ideal conditions

To break out in poison ivy after exposure

Pain to subside following a hemorrhoidectomy

A person to die of rabies after the onset of symptoms

For death to occur from Bubonic Plague

THREE TO SEVEN DAYS

For a felony defendant to have a preliminary hearing after presentment before a magistrate—*ideally*

FOUR DAYS

Maximum time that the opening of a lock may be delayed by the use of a time clock

For the antiepileptic drug trimethadione to reduce petit mal seizures from 70 per day to zero

The body to recover from jet lag—*it takes this period of time for body temperature cycles to adjust to a new day-night cycle*

A Mississippi song sparrow to build and line its nest

Ideal maximum time to provide counsel for a misdemeanor defendant after arrest

The period that the common cold is communicable—*two days before to two days after symptoms appear*

Period that life can be sustained on an artificial heart pump

The mother red deer to get around to feeding her young for the first time

Fever associated with measles to subside—*rubella*

A fertilized human ovum to grow to ninety cells

The life span of human blood platelets

To build a large concentric yurt—*The structures have been adapted from traditional designs of Mongolian herdsmen for economical housing by Dr. William S. Coperthwaite of Bucks Harbor, Maine*

To sprout chia seeds, flax seeds, or soy beans

Maximum time that fresh corn will remain flavorful in the refrigerator

The cut life of roses with preservatives (see also 2d)

The estrus cycle of the mouse

To sprout almonds, barley, garden cress, lentils, millet, mung beans, oats, rice, rye, and wheat

The period a kitten's eyes remain closed after birth

4.46d
Length of school suspension for black children in U.S. public schools—*Children's Defense Fund*

4.5d
A baby chick to develop in the egg: stage 3: elbow and knee joints developed

4.52d
The satellite Rhea to make one revolution around Saturn

110h
The amount of playing time obtained from standard silver oxide, hearing-aid batteries on an ultraminiature ear radio—*by Edmund Scientific called the Stick-It-In-Our-Ear-Radio*

112h
The mean interval between incidents of married coitus experienced by men and women 45 to 54 years old—*Morton Hunt study for Playboy*

4d, 21h
The average American male to return to work following illness

FOUR TO FIVE DAYS

The process of vinous fermentation to take place in the making of red wine

For an infant to coordinate movement of his arms in a defense reflex—*study by John Watson*

The eastern goldfinch to build a nest—*This nest will last for several seasons*

To recover from a severe influenza attack

FOUR TO SIX DAYS

To drain the ear following a mastoidectomy

The symptoms of amoebic dysentery to subside after administering emetine therapy

FOUR TO SEVEN DAYS

The incubation period of roseola infantum

FOUR TO EIGHT DAYS

Adults to recover from a mild attack of bacillary dysentery

FOUR TO TWELVE DAYS

Cotton seed to germinate under ideal conditions

FOUR TO FOURTEEN DAYS

The seed of watermelon, safflower, and dahlia to germinate under ideal conditions

FIVE DAYS

A baby chick to develop in the egg: stage 4: differentiation of first three toes

Programs are broadcast over American television 120 hours per week

The annual rosary pilgrimage to Lourdes, France— *October 3–7*

The larvae stage of the worker honeybee

Brazil's lenten carnival—*February 8–12*

The cut life of carnations with preservatives (see also 3d)

A transatlantic crossing on a cruise ship

Between Yom Kippur and the feast of Ingathering

All the golfers in the United States to complete an 18-hole game if they all head for the 9,926 golf courses in the U.S. at the same time—*based on an estimated 9,500,000 golfers and a 12-hour playing day*

First class mail delivery between East and West Coasts via U.S. Postal Service

The interval between saturation of average sandy soil to a depth of 1 inch for optimal growth of lawn grass

Waiting period between purchase and delivery of firearms in California

Most fish and amphibians to shed their eggs—*the rest of the year is spent in acquiring a new stock*

Time off for good behavior granted for each month of a term served for a federal offense if the sentence is not less than 6 months and not more than 1 year

To climb Mount Kilimanjaro

To protect someone exposed to streptococcal infection with penicillin or bacterial pneumonia with penicillin G

Between courses of one kind of oral sequential contraceptives (see also 7d)—*five days after the onset of menstrual flow*

United Parcel Service guarantee of coast-to-coast delivery—*business days*

A week according to the U.S.S.R. calendar of 1929

The period that a child with measles should be isolated after the rash appears

A baby chick to develop in the egg: stage 5: beak formed

Bedbug eggs to hatch after fertilization

Fermentation process in the making of commercial ale

To extract alcohol from potatoes

5d, 17m, 10s
Robert and Joan Wallick to fly around the world in a light plane—*1966 world record holders*

5.1d
The average American to return to work following an illness

126h
To lift 1,900 cubic yards of rock out of a blasted area 80-feet square using two, four-foot lifts

5.5d
The average American female to return to work following illness

5.88d
The satellite Triton to make one revolution around Neptune

5.9d
Average female student to return to school following illness

FIVE TO TWELVE DAYS

To cure gonorrhea with penicillin—*until 48 hours after fever subsides*

FIVE TO SEVENTEEN DAYS

Battery life of radio direction-finders used by ships—*based on four hours use a day*

SIX DAYS

Shipwreck victims to die of exposure at sea—*World War II studies of U.S. navy showed that 50% of those lost at sea had died by the sixth day in a life raft if the temperature dropped below 41° F.*

The second best time to get steamship reservations before sailing—*according to an old saying, "if you can't get space 6 months before sailing, you can 6 days before sailing" (presumably because of cancellations)*

Cocoa beans to ferment

The hexaemeron—the time it took God to create the universe—*according to the Judeo-Christian tradition*

The maximum incubation of yellow fever

A county fair—*most run Monday through Saturday, the longest practical period for keeping animals confined and produce fresh*

Time off for good behavior—*this is the time granted for each month of a term served for a federal offense if the sentence is more than a year and less than three years*

Postriders to make a round trip from Boston to Philadelphia on horseback when Ben Franklin was postmaster general

Secondary bacterial pneumonia to be suspected as a complication of influenza if fever has not subsided

An oral dose of ethambutol to be excreted by the body—*Ethambutol is used in the treatment of tuberculosis*

William Saroyan to write the play *The Time of Your Life*—*May, 1939*

Headache symptoms to appear after the administration of a spinal anesthetic

Sage wine to ripen in the cask

Leukemia drug therapy with mercaptopurine to lower the leukocyte count of a patient

6d, 1h, 36m
The average interval between incidents of masturbations among males 30 years old—*Morton Hunt study for Playboy*

6d, 5h
U.S. *Apollo VIII* to make the first manned voyage around the moon beginning December 21, 1968

6d, 8h
A day on Pluto—*compared to an Earth day*

6d, 13h
A Soviet worker to earn enough to buy a man's suit —*Sylvia Porter, 1973*

6d, 19h, 12m
The period of hospital confinement for surgical procedures among men 17 to 24 years old

6.39d
Pluto to make one rotation on its axis

SIX TO EIGHT DAYS

A crew of four or five men to frame-in the average tract house

ONE WEEK

The salt level of Carmen Island Salt Lake in Mexico to be restored after all the salt in the lake has been removed

The waiting period for a marriage license in Oregon

Newborn guppies to double in size

To lose a pound—*if caloric intake is restricted to 500 calories a day less than is needed to maintain an individual's energy demands*

Milk to spoil after "pull" date has expired

To "dry out" an alcoholic—*until withdrawal symptoms have passed*

Period after receipt of damaged baggage that an international airline passenger has to complain for a claim to be entertained

Thirty-two hundred contacts with prostitutes to occur in an American city with a population of 100,000

A mother bear to work up an appetite after coming out of hibernation

Implantation to occur in humans after the ovum has been fertilized

Joshua to surround and capture Jericho

Snakeweed to produce abortion in cattle after consumption of 20 pounds of the fresh plant

Danger from blister gas to remain after its introduction into a wooded area

Infant opossums to multiply their birth weight ten times—*at birth they weigh only 1/100 of a pound, 1/10,000 of their mother's weight*

The amount of time off for good behavior granted for each month of a term served for a Federal offense if the sentence is not less than three years and more than five years

Communicable period for chicken pox—*one day before rash appears until rash has crusted*

Homemade table beer to ripen in the cask before tapping

An appendectomy wound to heal

To cure illnesses by dancing in a trance state—*done among Betsileo tribesmen of Madagascar*

Time allotted for filing all motions and election of nonjury trial in misdemeanor cases after the appointment of counsel—*ABA minimum standards*

Isolation period for a child with pneumonia

Recommended period for curing the final coat of pliolite resin–based paint on swimming pools before filling the pool with water for the season

For chorionic gonadotropin to be secreted in humans following ovulation—*chorionic gonadotropin is a hormone of pregnancy*

A ring dove couple to build a nest

To grow a six-inch-high carpet of barley grass from seed under controlled conditions—*The Brookfield, Illinois, Zoo is using this approach to provide green diet supplement for many of its animals. The grass is sprayed with water for ten minutes every six hours and grown in trays under flourescent light. One pound of seed will yield 20 pounds of greens at the end of seven days.*

Initial period of violent fermentation in California wines

Newborn babies to wet or soil seventy to ninety diapers

Bacon to lose freshness in the refrigerator—*32°*

The hospital recovery period for the amputation of a toe

The mean interval between incidents of married coitus experienced by men and women 55 years old and over —*Morton Hunt study for Playboy*

Waiting period between purchase and delivery of firearms in Connecticut and Maryland if the carrier is not licensed

A bean weevil to die when exposed to temperatures of −6° C.

Safe period for consumption of refrigerated lunchmeat

Period following thyroid surgery before bathing is recommended

Roast beef to lose freshness in the refrigerator—*32°*

Minimum period of traction for a slipped disc

Onset of withdrawal symptoms in infants born to alcoholic mothers—*study at Georgetown University Hospital*

7d, 9h
A phase of the Moon—¼ *synodical month*

7d, 12h
Survival period for dogs given kidney transplants and
no immunosuppressive drugs (see also 23.7d)

7.9d
Subscribers of Blue Cross low-option insurance to
complete their average hospital stay (see also 9.8d)—
study of U.S. government workers and their families

SEVEN TO EIGHT DAYS

An addict to pass through the withdrawal stage of
long-lasting barbiturates—*hallucinations may persist
for several months thereafter*

SEVEN TO NINE DAYS

The yolk of a chicken egg to mature in the ovary

SEVEN TO TEN DAYS

The eye to heal after removal of a cyst

For the opening between left and right ventricles of
the heart to close in a baby after birth

SEVEN TO TWELVE DAYS

Dilated pupils to return to normal after a local application of the drug atropine

SEVEN TO FOURTEEN DAYS

Castor beans and tobacco seeds to germinate—*under ideal conditions*

1–2w
Children to recover from infectious hepatitis (see also 4–6w)

EIGHT DAYS

The Jewish holidays Chanukah, Sukkoth, and Passover

Wear life of average quality panty hose—Consumers Report Buying Guide

A freighter to sail from New York to the Panama Canal

Nineteenth-century playwright Louis Count D'Assas to write a five-act play—*before he died from food and sleep deprivation*

Salicylates to pass through the body—*substance related to aspirin, found in many artificial food colorings and preservatives and believed to cause hyperactivity in some children*

The vulva of the female flying squirrel to return to normal after heat is over

The amount of time off for good behavior granted for each month of a term served for a federal offense if the sentence is not less than five years and more than ten years

Cautious stair-climbing to be permitted following hernia surgery

To age raisin cider in bottles before drinking is recommended

Soybeans and kidney beans to germinate under ideal conditions

8d, 1h, 12m
The half-life of the radioisotope iodine 131

8.4d
Subscribers to Aetna low-option insurance to complete their average hospital stay (see also 10.3d)—*study of U.S. government workers and their families*

200h
The number of flying hours needed to gain a commercial pilot's license

NINE DAYS

Maximum time for harelip wound to heal following surgery

The estrus cycle of dogs

The critical period of smallpox

The Yeibichai—*The Navajo ritual of fall and winter in which Navajos impersonate the gods*

9.5d
Nestling period of young tree sparrows

9.8d
Average interval between incidents of intercourse experienced by single males 18 to 24 years old—*Morton Hunt study for Playboy*

Average interval between incidents of masturbations among females 18 to 24 years old—*Alfred C. Kinsey, University of Indiana study*

Subscriber to Blue Cross high-option insurance to complete an average hospital stay (see also 7.9d)—*study of U.S. government workers and their families*

TEN DAYS

A creditor's allowable period for returning a down payment on a rescinded contract (see also 3d)—*a provision of the Federal Reserve Truth in Lending law*

Nestling period for the Mississippi song sparrow before it attempts its first flight

Addictive doses of the hypnotic drug chloralhydrate to produce dermatitis

To cure cattle of liver flukes using the drug emetine —*hydrochloride*

Paramecium aurelia to reach a stable population size —*when grown in a laboratory*

Maximum for onion seeds to germinate under ideal conditions

For immunity to poliomyelitis to be insured after administering Salk or Sabin vaccine

The Standard Vehicle Company to turn out a custom-made, horse-drawn vehicle—*Standard Vehicle Company is one of the last of the buggy makers*

Parsnips to ferment in the process of making parsnip wine

To age dandelion winc

A freighter to sail from the port of Baltimore or New York to Lisbon

To root petunias, verbena, and torenia from cuttings

The annual Minneapolis Aquatennial—*a water carnival beginning the third Friday in July*

To cure a streptococcus infection with antibiotic ampicillin

Newborn monkeys to form strong attachments to their mothers by clinging to them

An egg to reach the age when it will stand on end in a pan of cold water (see also 3d)

To cure scarlet fever with penicillin G or erythromycin

Withdrawal from opiates with methadone

Quince to perspire after picking—*this is preparatory to mashing them for use in making wine*

A Navajo family to build a large hogan

The salable period of milk after pasteurization

The time savings to navigators traveling from the north Atlantic to the equator after the introduction of Maury's sailing charts—*printed in his* Physical Geography of the Sea *in 1854*

Period bandages must be worn following a cornea transplant

A broken nose to heal

A week according to the French revolutionary decimal calendar of 1795

Charge services such as American Express to bill customers after receiving charges from business establishments—*minimum time*

The cycle of yin and yang on the Chinese calendar

Maximum school suspension allowed by most boards of U.S. public schools

The human body to eliminate a dose of arsenic from the system

A decan—*in astrology, the period subordinate to the astrological sun sign*

To cure infants born with syphilis by using penicillin therapy

Period that a male moose stays with one female moose

Time off for good behavior granted for each month of a term served for a federal offense if the sentence is 10 years or more

A bill to become law if the President neither signs it nor vetoes it—*unless Congress adjourns*

To steep raisins for making raisin cider

Period to observe an animal for rabies after it bites someone

Thomas Edison and his staff at Menlo Park to invent something—*In Edison's words, "a minor invention every 10 days and a big thing every 6 months or so," —a goal he realized collecting over 400 patents in an 11-year period*

From Rosh Hashanah to Yom Kippur—*known as the Ten Days of Repentance*

A fever blister to heal

The earliest a person will exhibit symptoms of rabies after exposure (see also 30d; 50d; 1y)—*30 to 50 days is average*

10.3d
Subscribers to Aetna high-option insurance to complete their average hospital stay (see also 8.4d)—*study of U.S. government workers and their families*

1.5w
Average interval between contacts with prostitutes among American men in their thirties who engage in sexual relations with prostitutes—*Alfred C. Kinsey, University of Indiana study*

TEN TO FOURTEEN DAYS

A patient to resume normal sexual relations following a hemorrhoidectomy

Hospital recovery period for amputation of an arm or leg

TEN TO SIXTEEN DAYS

10–15d
For iodine therapy for hyperthyroidism to reach its maximum effectiveness

The turnover rate of human blood cells

10–16d

Pheasants to die from daily doses of the seed disinfectant casoron at concentrations of 1,189 milligrams per kilogram of body weight

TEN TO TWENTY DAYS

Period during which consumers are entitled to pay the net amount listed on their utility bills—*for gas and electricity*

The liver of a rat to regenerate

TEN TO THIRTY-SEVEN DAYS

10–30d

Fever associated with malaria to subside without treatment

10–31d

The period of impurity assigned to family members of a deceased according to Hindu castes: Brahman (10d); Kshatriya (12d); Vaisya (14d); Sudra (31d)—*the higher the caste, the shorter the time*

10–37d

The estrus cycle of the horse

ELEVEN DAYS

Young indigo buntings to get bold enough to try their first flight of 20 feet or more

11.5d

Nestling period of young chipping sparrows

11.7d
Period of hospital confinement for an oophorectomy and/or salpingectomy

Interval between sexual outlets among homosexual males 35 years old—*Alfred C. Kinsey, University of Indiana study*

11.9d
To count to 1 million at the rate of a number per second

TWELVE DAYS

The phase of follicle growth during human estrus

A chronic alcoholic to get over his hypersensitivity to alcohol after the alcohol antagonist disulfiram is administered

The half-life of bromides in the kidneys

The estrus cycle of the squirrel monkey

Lesser sandhill cranes to die from daily doses of DDT administered at a concentration of 1,000 milligrams per kilogram of body weight

Viola seeds to germinate under ideal conditions

Communicable period of German measles—*seven days before to five days after the rash appears*

Average number of days it rains in Copenhagen in April

A 3-foot layer of lava on the surface of the Earth to cool from 2,000° to 1,400° F.

A freighter to sail from the port of New York or Baltimore to Genoa

Maximum recorded time that a person can stay awake without suffering permanent damage—*study by Dr. Bernard L. Frankel, National Institutes of Mental Health*

Christmas—*traditionally December 25 to January 5 (Twelfth Night)*

Birthday celebration for the Virgin Mary in Coptic and Abyssinian churches—*it is celebrated on the first day of every month*

An August hurricane to run its course

The yellow-fever toxin to pass through the body of a mosquito

Period a patient must stay off his feet following surgery for a torn cartilage in the knee joint

12.5d

Polyphenylene oxide plastics used in autoclavable surgical tools to exhibit only a .75% "creep" under a 3,000-psi load—*Creep is the dimensional change of a plastic over time*

TWELVE TO SIXTEEN DAYS

Nestling period of young thrushes

THIRTEEN DAYS

Gestation of opossum

Male chicks to first attempt to copulate when they are given injections of anterior pituitary soon after hatching

Incubation of tit eggs

Interval between water saturation of clay soil to a depth of 2½ inches for optimal growth of lawn grass

The pupa stage of the worker honeybee

Opossum eggs to develop in the uterus

Incubation of the nightingale

13.2d
An American 65 years old or over to recover from an illness involving bed disability

13.46d
The satellite Oberon to make one revolution around Uranus

FOURTEEN DAYS

A fortnight

To see clinical improvement in tuberculosis patients after beginning antibiotic therapy

A skilled technician to thread a weaver's lace-making machine

The maximum milk should be stored in a home refrigerator

Showering to be permitted following vaginalplasty

The media to grow weary of the streaking fad—*Spring of 1974*

Average interval between masturbations among males at age 60 who have a record of high-frequency masturbation—*Alfred C. Kinsey, University of Indiana study*

Period that paté de foie gras remains fresh in the refrigerator

Easter lilies to turn brown if cut when partially open

Average interval between orgasms experienced by single women 18 to 24 years old—*Morton Hunt study for Playboy*

Whitney to complete a model of the cotton gin after conceiving the idea

Minimum to ripen brick cheese

An alcoholic to begin to experience hallucinations and delirium tremens while on a "binge"—*constantly intoxicated*

To ferment Bordeaux wines before they are racked off

To learn to use an artificial leg

After the birth of a worker honeybee for it to begin foraging

Surgical wounds of the small intestine to heal

A baby gorilla's eyes to focus after birth

To administer duck embryo or inactivated brain tissue vaccine to a person who has had mild exposure to rabies

Modern Bride magazine to accept or reject a manuscript

Lion cubs to gain full sight

Homemade beer to age so that it is drinkable if malt is made at 119° F.

Danger from blister gas to remain after introduction under winter conditions

Period that nuclear power reactors must be shut down so that the nuclear regulatory commission can inspect the reactors for cracks which might lead to pipe failures if the reactors' emergency systems have to be used

Interval between the two doses of pyruvinium given to cure pinworms

A couple to undergo an intensive training session for sexual dysfunction with Masters and Johnson

Before ovulation takes place in women after the onset of menstruation

A freighter to sail from New York to Los Angeles via the Panama Canal (see also 16d)

Incubation of the whooping crane

For the socket to heal sufficiently to be fitted with an artificial eye after surgery for eye removal

Incubation of German measles (rubella)

Adults to recover from a tonsillectomy

U.S. Army Reserve summer camp session

Maximum for petunia seeds to germinate under ideal conditions

To toughen up tomato seedlings before setting them out in the garden—*by gradually exposing them to more sunlight and cooler temperatures*

To make the movie *Deep Throat*

Interval between the last safe application of the pesticide Sevin and the harvest of leaf lettuce

14.4d
Average period of restricted activity for men suffering an arthritis attack

FOURTEEN DAYS TO EIGHT WEEKS

2–3w
The mating season of male ground squirrels

2–4w
Period before surgery during which women should refrain from taking birth control pills—*to avoid increased risk of postsurgical blood clots*

2–8w
The symptoms of schizophrenia to recur after cessation of the "taming" drug phenothiazine derivative

FIFTEEN DAYS

An organ transplant patient to be reasonably sure his body will not reject the transplant

Interval between washings of the 6,500 windows in the Empire State Building

Waiting period between purchase and delivery of firearms in Tennessee

A freighter to sail from the port of New York or Baltimore to Marseilles

Young house sparrows to learn to fly—*birth to 15 days*

To cure an attack of dumdum fever with antimony

2.16w
Average interval between incidents of masturbations among married men in their late twenties and thirties —*Alfred C. Kinsey, University of Indiana study*

SIXTEEN DAYS

Period that the Mormon temple in Utah was open for public tours—*Mormon temples are not open to the public after dedication ceremonies but the new interior of the building made it new in the eyes of churchmen and so it was opened after extensive renovation for a brief time in the Spring of 1975 before rededication.*

Period that mumps is communicable—*The virus causing the disease is present in the saliva up to seven days before and nine days after the swelling appears.*

The queen honeybee to progress through stages from egg to adulthood

16d, 16h
A Navajo weaver to make a 3-by-5-foot-rug—*90 threads to the inch*

To build the scale model of the NY-2 biplane in which Jimmy Doolittle performed the first instrument landing—*on view at the Smithsonian Institution*

SEVENTEEN DAYS

Doris Day to earn one million dollars cutting dog food commercials for General Foods

Average number of days it rains in London in January

Average American female to recover from an illness in which activity is restricted

EIGHTEEN DAYS

Most patients to die following a lung transplant

Mumps vaccine to produce immunity after innoculation

Average time to get a District of Columbia insurance commissioner to investigate a complaint on cancellation of an insurance policy

The incubation of the roadrunner

Seals to be weaned

The heart to begin forming in a human embryo

NINETEEN DAYS

Young cardinals to make their first flight

Gestation of the mouse

TWENTY DAYS

A week—*according to the ancient Aztec calendar*

A course of the oral contraceptive enovid—*days 5 to 25 of the menstrual cycle*

Polecat cubs to open their eyes

The siege of Yorktown by American and French forces—*The British surrendered October 19, 1781*

The Australian budgerigar to die without water—*parakeet*

500h
A diamond stylus on a turntable to wear out

20–30d
The period of egg-laying delay among vertebrates for each 10 degrees of latitude toward the poles

TWENTY-ONE DAYS

Interval between the world's first test detonation of an atomic bomb in New Mexico and the dropping of such a bomb on the civilian population at Hiroshima, Japan

Incubation of German measles (rubella)

Maximum time to get a passport after application

The rotting process which prepares flax fiber for weaving

To polish stones in an electric fruit jar rock tumbler

Maximum time for carrot seeds to germinate under ideal conditions

A grape to become a raisin in the California sun

Symptoms of sympathetic ophthalmia to begin after eye injury—*for the uninjured eye to begin to go blind after injury to the other*

The estrus cycle of goats

A wedding feast in Yemen

Minimum term of a Eurailpass allowing unlimited train travel in thirteen European countries—*others may be purchased for one, two, or three months*

Period of quarantine for the first men on the Moon in 1969

Period prior to surgery when a person can have up to 3 pints of blood drawn so that he can have transfusions of his own blood

A mature shipworm to bore through a 16-inch ship's timber

A freighter to travel from New York to Australia via Panama Canal

The wound to heal from a radical mastectomy

Lion cubs to get their first teeth

Maximum period of estrus in domestic cats

A worker honeybee to progress through the stages from egg to adulthood (see also 24d)

To cure and hickory smoke bacon

Limit placed on Congress to rule on the inability of a President to execute his duties—*provided by the 25th amendment to the Constitution*

Leukemia to exacerbate after withdrawal of chemotherapy with vinblastine

A silk worm to complete metamorphosis and emerge from the cocoon

Incubation of chicken pox

Grass and clover pasture to be renewed after cattle have grazed

To renew sexual relations following a hymenotomy

The caterpillar stage of the monarch butterfly

The average worker's body to adjust to a new work shift—*day to night shift affects the biological cycles governing sleep, metabolism, temperature, excretion, etc.*

TWENTY-TWO DAYS

Estrus cycle of the hanuman langur

Gestation of rats

22d, 7h, 56m
Gemini V to make the first extended, manned space flight beginning August 21, 1965

TWENTY-THREE DAYS

The human biorhythm cycle of physical strength, endurance, energy, resistance, and confidence (see also 28d and 33d)—*The theory, first postulated in Europe at the turn of the century, is again gaining prominence as airlines seek to halt a rash of pilot-error accidents.*

Beginning the individual's cycle at the date of birth, it is possible to predict "off" days or critical days when judgment may be impaired.

"Executive" monkeys to die from bleeding ulcers during testing where they are exposed to extended daily periods of making decisions under stress—*tests of the Walter Reed Army Institute of Research*

23.7d
Average survival period of dogs given kidney transplants and the immunosuppressive drug 6 mercuptopurine (see also 7.5d)—*but some live 8 to 10 years*

TWENTY-FOUR DAYS

Incubation of mumps

Average wait for unemployment benefits in the District of Columbia

The drone honeybee to progress through the stages from egg through adulthood (see also 21d)

TWENTY-FIVE DAYS

Handel to compose "The Messiah"

Average number of flood days in Venice between October and April

A mouse to come into her first heat—*although she will not reach full physical maturity until she is 50 days old*

Apparent sidereal rotation of the Sun

TWENTY-SIX DAYS

Average waiting time to get a gun license in the District of Columbia

An injection of penicillin G to be totally excreted from the body

TWENTY-SEVEN DAYS

Uterine gestation in spiny anteaters before the egg is deposited in a pouch and incubated there

Interval between nesting and egg laying for eastern goldfinches

27d, 7h, 43m, 11.5s
A sidereal month as compared to a solar calendar

TWENTY-EIGHT DAYS

The incubation of sea gulls

Rooting roses and evergreen shrubs from cuttings

Advent—*the period including four Sundays before Christmas in Christian churches*

Women to be permitted to resume sexual relations following childbirth or most obstetrical or gynecological surgical procedures

Maximum that ice cream should be stored in a home freezer

Concrete to cure to its greatest strength

Use of the eyes to be permitted following removal of a cataract

To train a blind person to use a seeing-eye dog

The human biorhythm cycle governing sensibility, nerves, feelings, intuition, cheerfulness, moodiness, and creative ability (see also 23d)

The witch to try to fatten up Hansel—*in the Grimms' fairy tale*

Elephant seals to molt

Life span of a worker honeybee born in midsummer (see also 6mo)

The brain of a rat to reach 80% of its mature size

Gestation of the rabbit

The human embryo to develop blood vessels connected to the heart

Male guppies to reach adult size (see also 12w)

For dancing to be permitted following an appendectomy

To cure staphylococcal infections with antibiotics

February—*except during Leap Year*

The human ovarian cycle

Users of contact lenses to build up a tolerance to the "hard type" lenses by wearing them only 2 hours a day at first and gradually increasing wear time

TWENTY-NINE DAYS

February—*every Leap Year*

To use up current world-wide food reserves

Number of days under the sign called Capricorn (the goat) in the Zodiac—*December 22 to January 19*

700–900h
Average life of a mercury battery used in smoke detectors

29d, 12h, 44m, 3s
From new Moon to new Moon—*the synodical months*

THIRTY DAYS

April, June, September, and November

The maximum that chiffon pie, frosted cake, cooked leftovers, and sandwiches should be kept frozen

For 25% of women in their late teens or twenties to get pregnant when they have intercourse every other day

The "grace period" on most life insurance policies—*the period following the due date when premiums can be paid before penalty or cancellation of policy*

Life span of head lice

A tourist visa in Afghanistan, Costa Rica, Haiti, and Hungary

Beaver's teeth to grow at a rate of 1 inch during its adult life—*They are worn down at an almost equal rate by gnawing wood.*

Group medical plans to be cancelled if an employee is laid off—*most can be converted to individual policies with fewer benefits for a higher premium*

The symptoms of Bang's disease to manifest themselves in dairy cattle after abortion or birth

To stabilize solid waste by retention in wastewater treatment plants

The period a plastic-lined swimming pool should cure before painting it for the first time

An infant to be able to hold objects for 30 seconds or more

Maximum length of each of two periods of confinement that is covered by Blue Cross for treatment of alcoholism

One fourth of the population in Bombay to earn $27

Quarantine period for birds entering the United States

Between peggings of the wine cask during ripening to check on the clarity of the wine

Between four and five thousand people to request documents at the National Archives in Washington, D.C.

An infant to become capable of a regular flow of tears—*study by John Watson*

Maximum reasonable period for preparation of transcripts after the close of a trial—*ABA minimum standards*

Human hair to grow an average of a half inch

An infant to learn voluntary smiling—*study by John Watson*

The human body to reach the limit of its tolerance to 5 roentgens of protracted radiation exposure a day

Baby alligators to grow about an inch

Radishes to mature

State residency requirement for voting—*A 1972 Supreme Court ruling prohibits the longer waiting periods imposed by many states.*

To cure green unpeeled fence posts after soaking in zinc chloride before setting them in the ground

Period a race horse bought in a claiming race must be held before it can be sold—*except through another claiming race*

The number of days included under the following signs of the Zodiac: Aquarius, the water bearer (January 20 to February 18); Pisces, the fish (February 19 to March 20); Taurus, the bull (April 21 to May 20); Libra, balance (September 23 to October 22); Scorpio, the scorpion (October 23 to November 21); and Sagittarius, the archer (November 22 to December 21)

The female guinea pig to reach breeding age

To lose 4 pounds of body weight if caloric intake is restricted to 500 calories less than is needed to maintain an individual's daily energy demands

To begin to detect cystic fibrosis in children

Average time for onset of symptoms of rabies after exposure (see also 10d; 50d; 1y)—*30 to 50 days is average*

To extinguish an Alaskan forest fire—*fire penetrates the permafrost which, like a peat bog, keeps fueling the flames*

THIRTY TO FIFTY-FOUR DAYS

To survive dog days each year—*Usually between July 3 and August 15, known for their hot, sultry nature in the northern hemisphere and named for Sirius.*

THIRTY TO SIXTY DAYS

Periodical publishers to process subscribers' changes of address

THIRTY-ONE DAYS

January, March, May, July, August, October, and December

The survival period for one third of the patients who undergo bone marrow transplants

Minimum interval for selling and buying shares of the same stock under federal law

The estrus cycle of the macaque

The number of days included under the following signs of the Zodiac: Aries, the ram (March 21 to April 20); Gemini, the twins (May 21 to June 21); Cancer, the crab (June 22 to July 22); Leo, the lion (July 23 to August 22); and Virgo, the virgin (August 23 to September 22)

THIRTY-TWO DAYS

Mercerized cotton cord to lose 50% of its strength when heated in a test oven at 130° C. (266° F.) and untreated cotton cord to lose 60% of its strength

THIRTY-THREE DAYS

To reach lag b'Omer during the counting of Omer— *Jewish holiday*

The human biorhythm cycle governing intelligence, memory, mental alertness, logic, reasoning power, reaction, and ambition (see also 23d)

Period a baby polar bear's eyes remain closed after birth

THIRTY-FOUR DAYS

Americans 65 years old and over to recover from an illness in which activity is restricted

THIRTY-FIVE DAYS

A mouse to reach sexual maturity

The summer music festival in Salzburg, Austria— *home of Wolfgang Amadeus Mozart*

Gestation of the koala bear

Cress to mature

A kitten to acquire all its milk teeth

The incubation of the southern bald eagle

To become an airline hostess

THIRTY-SIX DAYS

The limit of regular sessions of the state legislature in Alaska and Arizona

The estrus cycle of the chimpanzee

THIRTY-EIGHT DAYS

Passage on a slow boat to China—*a freighter with passenger accommodations from New York to Hong Kong*

Review time for nonemergency legislation presented by the District of Columbia City Council to Congress (see also 58d)

THIRTY-NINE DAYS

A human fetus to reach the stage when doses of thalidomide cause shortening of the limbs

The estrus cycle of the gorilla

The gestation of the kangaroo

FORTY DAYS

The period of fasting at Lent (Ash Wednesday to Easter) required in Christian churches—*46 calendar days in the Western church; 56 in the Eastern church*

The period mourning men in parts of Bulgaria refrain from shaving following the death of a loved one

The period between Christ's resurrection and ascension to heaven

The flood to reach its peak after Noah was safely in the ark—*Genesis 7:17*

The average number of days lost from work by those suffering whiplash

The period after birth when Arab infants wear swaddling clothes

Mustard to mature

The Norway rat to reach sexual maturity

FORTY-ONE DAYS

The incubation of turkey vulture eggs

41d, 16h
The life span of a tungsten filament lamp

Classroom and on-the-job training time for qualified masseuses at an accredited academy—*law in Falls Church, Virginia, designed to close down massage parlors*

FORTY-TWO DAYS

To steep cherries in brandy before bottling homemade black cherry brandy

Period a patient is restricted from driving a car following a cornea transplant

The honeymoon of striped bass

Watergate conspirator, John W. Dean to gross more than $100,000 on a college lecture tour in the Spring of 1975—*He began the tour less than one month after being released from prison, having served four months of a one-to-four year sentence for conspiracy to obstruct justice.*

To become a notary public

A patient to be restricted from drinking alcoholic beverages following removal of a cataract

The amount of time Bluebeard says he will be away when he surprises his wife after she has opened the forbidden door

Period to form or break a habit—*according to Jackie Rogers, psychologist and founder of Smokenders*

Incubation of the golden eagle

Maximum time IRS has to mail tax refunds to citizens if they make the April 15 filing deadline—*after May 31, the agency must pay 7 percent interest on any money owed*

A woman to recover from ovarian surgery so that she can resume normal activities including driving and housework

Period of additional hibernation for the groundhog if he sees his shadow on February 2

The average period of remission in multiple sclerosis patients

The aging of Chinese "thousand-year" eggs

Basic training in the U.S. Air Force

The raising of veal for slaughter

The recovery period for whooping cough

The minimum dry period in dairy cattle before calving and the period it takes to reach the peak of milk production after birth

Positive confirmation of human pregnancy after conception

Collarbones and wristbones to knit following fracture

The estrus cycle of the slender loris

FORTY-THREE DAYS

The incubation of eagles

FORTY-FOUR DAYS

Martinmas—*the period of fasting in the Medieval Christian Church from November 11 to Christmas— St. Martin's Lent*

The testing period for new watches—*the Kew-Teddington Observatory*

The gestation of the squirrel

FORTY-FIVE DAYS

The limit on regular sessions of the state legislature in Georgia and South Dakota

The nestling period of ospreys

Suckling period for rats

FORTY-SIX DAYS

46d, 7h, 12m
The half-life of the radioisotope iron 59 (see also 2y, 329d)

FORTY-EIGHT DAYS

The gestation of mink

FORTY-NINE DAYS

Between Passover and Shavuot—*known as the counting of the Omer*

Buddha to achieve nirvana while meditating under the bo tree

To recover from diphtheria

The gestation of the skunk

Between Easter and Whitmonday—*French holiday*

Gestation of the fox

FIFTY DAYS

The incubation of cuttlefish eggs

Nestling period of young buzzards

FIFTY-ONE DAYS

A bean weevil to die when exposed to temperatures of 11° C.

To use up a 1½-ounce bottle of Ban roll-on deodorant —*fourteen days longer, the company claims, than the leading 5- and 6-ounce antiperspirant sprays*

FIFTY-FOUR DAYS

The Turkish siege of Constantinople—*The city surrendered May 29, 1453.*

FIFTY-FIVE DAYS

The Pilgrims to cross the Atlantic aboard the *Mayflower*

Between the flowering of cotton and the opening of the cotton boll

FIFTY-SIX DAYS

Maximum harvesting period for mature asparagus plants—*at least 4 years old*

Children 6 months to 6 years old to outgrow their shoes

The period during which a newborn meadow mouse gains one half its birth weight every single day

Basic training in the U.S. Army Reserve

Toes and fingers to be first visible on a human fetus

Young cardinals to strike out on their own

A human embryo to become a fetus—*recognizable organism*

To resume sexual relations following prostate surgery

To ripen bel paese—*an Italian cheese*

Tweezed hairs to resurface on the skin

To ripen port du salut cheese—*made by Trappist monks in France and Canada*

FIFTY-SEVEN DAYS

Dairy cattle to go into heat after calving (see also 75d and 60–90d)

FIFTY-EIGHT DAYS

Congress to act on the first twelve nonemergency bills presented by members of the City Council of the District of Columbia in 1976 (see also 38d)

58d, 8h
To customize a Jaguar 3.5 sedan by covering the entire body with redwood which is protected with ten coats of lacquer—*Jack Wood of Monterey, California*

58.65d
Mercury to make one rotation on its axis

SIXTY DAYS

Golden hamsters to reach sexual maturity

The drug emetine to be eliminated from the body—
Emetine is used to cure dysentery

The human fetus to develop fat tissue under the skin

Sycamore wine to ripen in the cask

To ripen the Italian cheese asiago for table use

To drive cattle up the Chisholm Trail from Texas to
Abilene, Kansas, for slaughter—*before the introduction of refrigerator cars on trains*

To become dependent on barbiturates to the extent
that denial causes seizures and delirium tremens—
*based on continual heavy intoxication during this
period*

Infant moles to reach the size of their parents

Mallard ducks to acquire flight feathers

To hand carve a dotara—*a two-stringed mandolin
found in India*

Young turkey vultures to reach full development

Infants to reach the age when they are usually given
their first combined diphtheria-tetanus-pertussis
(whooping cough) shot and first oral polio vaccine

After a caesarean section for a woman to be permitted
to resume all normal activities

The limit on regular sessions of the state legislature
in Arkansas, Florida, Hawaii, Idaho, Kansas, Kentucky, Louisiana, Montana, New Mexico, North
Dakota, Rhode Island, Utah, Virginia, Washington,
and West Virginia

The possessions of a deceased person in New South Wales to be decontaminated by hanging them in a tree

The period of daylight beginning June 21 at 70 degrees latitude in the northern hemisphere and on December 21 at 70 degrees latitude in the southern hemisphere

Average duration of morning sickness during early pregnancy

Suckling period for cats

Minimum aging that makes homemade wines fit to drink

Peas to mature

To ripen Dutch Edam cheese

The estrus cycle of the platypus

Period of relief cortisone injection brings arthritis patients—*maximum time*

A freighter to sail from New York to Cape Town, South Africa and return

Bush beans to mature

Most meats and fruits to lose their freshness when stored in a freezer at 10° F. (see also 1y)

Social Security and Medicare payments to begin after notification

SIXTY TO NINETY DAYS

Sweet corn to mature

SIXTY TO ONE HUNDRED DAYS

To fatten cattle for slaughter

Minimum time for relapses of dumdum fever to occur

SIXTY TO THREE HUNDRED SIXTY DAYS

To ripen American cheddar or Italian provolone cheese

SIXTY-ONE DAYS

The male indigo bunting to average 263,520 songs

The limit on regular sessions of the state legislature in Indiana

The average gestation for domestic dogs

SIXTY-TWO DAYS

1,500h
To gain the flight experience to earn an airline transport license

Average burn life of incandescent light bulb

SIXTY-THREE DAYS

The gestation of the wolf

Duration of Smokenders course

Baby gorillas to begin crawling

The gestation of the domestic cat

The gestation of the otter

One thousand broiler chickens to produce 2,700 pounds of manure—*four pounds per bird*

To produce a 3-pound broiler chicken

The gestation of guinea pigs

Octopus eggs to hatch after fertilization

SIXTY-FOUR DAYS

Incubation of the emperor penguin

SEVENTY DAYS

To make a piece of Wedgwood china

Hysterectomy patient to be able to resume all activities including athletics and any heavy work following surgery

The nestling period of the golden eagle

A baby polar bear to acquire acute hearing

The limit on regular sessions of the state legislature in Maryland

A clawed jird to reach sexual maturity

A salamander's leg to regenerate

Period a cast is worn following fracture of the upper end of the shinbone

SEVENTY-TWO DAYS

Cherry tomatoes to mature

SEVENTY-FOUR DAYS

74d, 12h
The half-life of the radioisotope iridium 192

SEVENTY-FIVE DAYS

The incubation of the kiwi bird

A cow to go into heat after birth of her first calf

A chipmunk to reach sexual maturity

A bedbug to lay 150 eggs

SEVENTY-SIX DAYS

The time saving to navigators sailing from the North
Atlantic to California after the introduction of Maury's
sailing charts—*printed in his* Physical Geography of
the Sea *in 1854*

SEVENTY-NINE DAYS

79.33d
The satellite Iapetus to make one revolution around
Saturn

EIGHTY DAYS

The incubation of the albatross

The period of suckling in a mother opossum's pouch

From the time of planting cotton to its flowering

EIGHTY-FOUR DAYS

Period for wearing a cast following fracture of the heelbone—*os calcis*

Female guppies to reach adult size (see also 4w)

A family of northern bald eagles to tread their eyrie flat

Children 6 to 10 years old to outgrow their shoes

EIGHTY-FIVE DAYS

The half-life of scandium

EIGHTY-SEVEN DAYS

87.96d
A day on Mercury—*Compared to Earth days. Note: Mercury's day and year are of the same length since it makes only one rotation during a year; one side faces the sun continually.*

EIGHTY-NINE DAYS

The American clipper ship *Flying Cloud* to sail from New York to San Francisco

89d, 1h
Winter—*from the winter solstice to the vernal equinox*

89d, 16h
Autumn—*from the autumnal equinox to the winter solstice*

NINETY DAYS

The period Aladdin was to have waited for the hand of the sultan's daughter

Compost to ripen

The waiting perod for a divorced Moslem wife before remarrying—*this insures that she is not pregnant by her former husband*

The minimum period after the first late mortgage payment notice that a home mortgage can begin foreclosure procedures

The maximum time smoked sausage, shrimp, and hot dogs should be kept in the freezer

Period a cast is worn following fracture of the midshaft of the shinbone

The suckling period of fur seals

The minimum aging time for the following cheeses: Swiss, gruyere, Italian gorgonzola, Italian caciocavallo, and blue cheese

The number of frost-free days necessary for wheat to mature

The human fetus to reach the peak hazard period for German measles

The average wait for billboard space

The period during which Coors beer distributors are allowed to sell a batch of the famous Colorado brew —*they are then instructed to destroy it*

Avondale shipyard to build a ship

Dairy cattle to wean calves

The current average hospital confinement for tuberculosis (see also 7mo and 10mo)

Homemade beer to age so that it is drinkable if malt is made at 129° F.

To run out of Medicare during any one benefit period —*does not include 60-day lifetime reserve*

Turnip wine to ripen in the cask before bottling

To be assured syphilis will not recur after ending penicillin therapy—*Serological tests must be negative during this period.*

Early potatoes to mature

The limit on regular sessions of the state legislature in Delaware, Oklahoma, and Tennessee

Okra and late beets to mature

To determine the cure of beef tapeworm

For pregnancy to be advisable after a myomectomy— *earliest date*

Delivery of sea mail between continents of Africa and North America

Period a canned ham remains fresh in the refrigerator —*unopened at 32°*

Spinach to mature

The gestation of the puma

To lager commercial beer

An exceptional, 10-year-old fur-seal bull to service 150 cows during a single mating season

To train as an intercity bus driver

For the human bile ducts to take over the efficient functions of the gall bladder following its surgical removal

To resume meals of normal size following stomach surgery

Leeks to mature

Earliest a football player can resume play following an operation for torn cartilage in the knee joint

THREE TO FIVE MONTHS

To raise "spring" lamb for slaughter

THREE TO SIX MONTHS

To raise turkeys for slaughter

Bathing to be permitted following spinal fusion surgery

NINETY-ONE TO NINETY-NINE DAYS

91d, 7h, 26m, 24s
The Earth to fall into the Sun if the Earth should leave its orbit

92d, 23h
Spring—*from the vernal equinox to the summer solstice*

93d
A flicker to continue to lay eggs if her clutch is kept small

93d, 13h
Summer—*from the summer solstice to the autumnal equinox*

98d
To become a G man—*The period of training at the Quantico, Virginia FBI Academy.*

Two-and-a-half million French men and women to see the film *Emmanuelle* as soon as it was released

ONE HUNDRED DAYS

Period within which the human body will normally reject a transplanted organ

To build the U.S. *Monitor* during the Civil War—*the ironclad ship that influenced a century of warships*

A snapping turtle in Mansfield, Ohio, to correctly predict the weather 85 times—*by sitting in the portion of his aquarium marked "Fair," "Changeable," or "Rainstorm"*

Early carrots to mature

ONE HUNDRED FIVE DAYS

Bears to hibernate

An ass to die from whitebrush (*Aloysia lyseiodes*) poisoning

The growing season in Manitoba

The gestation of the lion and the tiger

ONE HUNDRED TWELVE DAYS

A thousand broiler turkeys to produce 4,320 pounds of manure—*8 pounds per bird*

The gestation of the pig

ONE HUNDRED FIFTEEN DAYS

The gestation of the ocelot

ONE HUNDRED SIXTEEN DAYS

The fingernail of a young man to grow from the cuticle to the end of the nail—*from a personal study by Dr. William B. Bean at the Institute for Humanities in Medicine, University of Texas at Galveston, a growth of .45 centimeters*

The synodic period of Mercury

ONE HUNDRED SEVENTEEN DAYS

The body to replace a lost fingernail or toenail

ONE HUNDRED NINETEEN DAYS

The summer residency in Manitoba of the ruby-throated hummingbird

Average period Australian immigrants spend at hostels —*they are entitled to stay as long as one year*

ONE HUNDRED TWENTY DAYS

Life span of human blood cells

Gestation of the giant panda

Cats to cut their permanent teeth

ONE HUNDRED TWENTY DAYS TO ONE HUNDRED FIFTY DAYS

Pumpkins, watermelon, cauliflower, and cabbage to mature

ONE HUNDRED TWENTY-ONE TO ONE HUNDRED TWENTY-NINE DAYS

121d
The half-life of the radioisotope thulium 170

125d
Life span of human red blood corpuscles

126d
The siege of the British garrison at Kimberley by the Boers—*The garrison was relieved on February 15, 1900.*

Growing season in Cheyenne, Wyoming

129d
The growing season in Concord, New Hampshire

ONE HUNDRED THIRTY DAYS

Period a Saudi Arabian widow must wait before remarrying

Maximum period of unemployment acceptable to government in USSR—*remain without a job any longer than that and you are shipped off to a labor camp and labeled a "social parasite"*

Annual visit of the 120-day wind of Seistan Basin, Iran

To train a seeing-eye dog

Infants to reach the age when they are usually given their second combined diphtheria-tetanus pertussis (whooping cough) shot and second oral polio vaccine

The upper canine teeth of the baby walrus to erupt —*tusks*

Late carrots to mature

A normal red blood cell to rupture

The longest most fish should be kept in the freezer

The limit on regular sessions of the state legislature in Minnesota and North Carolina

Dairy cattle to produce half their annual production of milk after the onset of lactation

The American worker each year to work off his portion of federal, state and local taxes owed

ONE HUNDRED THIRTY-ONE TO
ONE HUNDRED THIRTY-NINE DAYS

131d
The British siege of Delhi—*Indian mutineers surrendered September 20, 1857.*

132d
Seasonal residency of marsh hawks in Montreal

133d
Wintering period of marsh hawks aday from Montreal

134d
The growing season in Great Falls, Montana

135d
The siege of Paris by the Germans—*The city surrendered January 28, 1871.*

137d
The growing season in Portland, Maine, and Sault Saint Marie, Michigan

ONE HUNDRED FORTY DAYS

The growing season in Reno, Nevada

The limit on regular sessions of the state legislature in Texas

Folic acid deprivation to produce anemia

ONE HUNDRED FORTY TO
ONE HUNDRED FORTY-NINE DAYS

144d
The summer residency of the eastern night hawk in Rochester, New York (see also 178d)

146d
To travel from Earth to Venus

148d
The fingernail of a middle-aged person to grow from the cuticle to the end of the nail (see also 116d)

ONE HUNDRED FIFTY DAYS

To fatten cattle for slaughter

Late filers of income tax to incur a penalty of 25% of their tax liability—*For those later filers who have not gone through the proper extension procedures (5% for each month late up to 5 months).*

Infants to lose the grasping reflex

The growing season in Sioux Falls, South Dakota

The limit on regular sessions of the state legislature in Connecticut

Flood waters to abate after Noah was safely in the ark (see also 40d)—*Genesis 8:3*

The gestation of sheep

Minimum to ripen Swiss sap sago cheese

Celery and brussels sprouts to mature

FIVE TO SIX MONTHS

Bordeaux wines to mature

ONE HUNDRED FIFTY-ONE TO
ONE HUNDRED FIFTY-NINE DAYS

154d
Hens to reach laying age

155d
The growing season in Burlington, Vermont

Between visits to the dentist among children of middle-
and upper-income families (see also 1y, 90d)—*Mollie
and Russell Smart*, Children, Development and Relationships

158d
The growing season in Boise, Idaho

ONE HUNDRED SIXTY DAYS

The gestation of the macaque

ONE HUNDRED SIXTY-ONE TO
ONE HUNDRED SIXTY-NINE DAYS

164d
The growing season in Denver, Colorado

165d
Rabbits born in the fall to come into their first heat
(see also 8mo, 15d)

166d
The growing season in Minneapolis–Saint Paul

168d
The growing season in Albany, New York

One thousand heavy turkeys to produce 35,000 pounds
of manure—*20 pounds per bird*

ONE HUNDRED SEVENTY-ONE TO
ONE HUNDRED SEVENTY-NINE DAYS

172d
The growing season in Des Moines, Iowa

174d
The growing season in Spokane, Washington, and
Wichita, Kansas

175d
The growing season in Hartford, Connecticut

176d
The growing season in Milwaukee, Wisconsin

177d
The growing season in Buffalo, New York

178d
The summer residency of the eastern night hawk in
New Orleans (see also 144d)

ONE HUNDRED EIGHTY DAYS

To cure hardwood for fireplace use

Tay-Sachs disease to be detectable in children of Jewish parents

Lion cubs to become mature enough to join in the hunt

Minimum wait to have a U.S. patent granted (see also 2y)

The transition period for phasing out services to outgoing U.S. presidents

Prison term for a person convicted of violating immigration laws

A human infant to grow to the extent that his thorax is larger than his head

Whiplash symptoms to persist (see also 40d)

Minimum period for aging a Smithfield ham

A human infant to reach the peak of bodily heat production for his entire life—*heat production per unit of body weight*

To replenish body stores of iron with oral iron therapy in patients suffering from iron deficiency anemia

The weight curve of infants in impoverished countries to drop below normal even when nursing continues for several years

A young aardvark to gain maturity needed to dig its own burrow and fend for itself

Life span of a worker honeybee born in late summer
(see also 4w)

Valid period for a prescription for the tranquilizers
valium and librium—*The FDA curbs allowing no
more than five refills during this time, began in March
1975.*

Gestation of the baboon

Cowslip mead to ripen in the cask

Maximum period pork and oysters should be kept in
the freezer

An infant to begin to need more nutrients than can
be provided by mother's milk

Cattle to reach sexual maturity

A nursing mother to use up her own stores of iron
without supplement

An infant to surpass the critical period for occurrences
of crib death

Time subtracted from the maximum prison sentence
for purposes of determining the length of parole

A human fetus to develop so that it can survive if
born prematurely

The length of the school year in most U.S. school
districts

To complete the Marine Corps infantry officers Basic
school at Quantico, Virginia

Average playing season of the major symphony or-
chestras in the United States

ONE HUNDRED EIGHTY-ONE TO
ONE HUNDRED EIGHTY-NINE DAYS

183d
Seasonal residency of Maynard's cuckoo in Florida

185d
A compulsory school year in Arizona

186d
Siege of Vicksburg by federal troops during the U.S. Civil War—*The city surrendered July 4, 1863.*

187d
The growing season in Louisville, Kentucky

ONE HUNDRED NINETY DAYS

Lifetime mental hospital benefits under Medicare

The growing season in Detroit, Michigan, and Wilmington, Delaware

ONE HUNDRED NINETY-ONE TO
ONE HUNDRED NINETY-NINE DAYS

192d
The growing season in Cincinnati, Ohio, and Providence, Rhode Island

193d
The growing season in Baltimore, Maryland

194d
The growing season in Cleveland, Ohio

195d
The limit on regular sessions of the state legislature in Missouri

The growing season in Albuquerque, New Mexico

TWO HUNDRED ONE TO TWO HUNDRED NINE DAYS

202d
The growing season in Salt Lake City, Utah

203d
A school year in the Soviet Union

205d
The growing season in Saint Louis, Missouri

206d
The growing season in Kansas City, Missouri

208d, 8h
Burn life of a fluorescent lamp

TWO HUNDRED TEN DAYS

California courts to award more than one million dollars in malpractice suits to plaintiffs at the rate of one a month—*Such experiences in the past several years have affected the cost of malpractice insurance tremendously and in turn the cost of patient care.*

Gestation of the black bear

Maximum length of a session of the U.S. Congress—*January 3 to July 31 except in time of war*

Gestation of the white-tailed deer

Breeding season for female cotton rats in northern latitudes (see also 9mo and 12mo)

The male offspring of a weasel to grow larger than his mother

To wean a blue whale

A female plant louse of the species *Phylloxera vasta-trix* to lead a reproductive cycle resulting in 25 million descendants—*This species which feeds on the roots of grape vines was responsible for devastating European vineyards in the 1870s.*

**TWO HUNDRED ELEVEN TO
TWO HUNDRED NINETEEN DAYS**

215d
The growing season in Pittsburgh, Pennsylvania

216d
The growing season in Boston, Massachusetts

218d
The growing season in New York City and Omaha, Nebraska

219d
Lifetime sleep requirement of a cow—*A cow sleeps only three percent of the time.*

The growing season in Richmond, Virginia

TWO HUNDRED TWENTY DAYS

Chimney swifts to winter away from Madison, Wisconsin

TWO HUNDRED TWENTY-ONE TO TWO HUNDRED TWENTY-NINE DAYS

223d
The growing season in Nashville, Tennessee, and Oklahoma City, Oklahoma

224d
The growing season in Atlantic City, New Jersey

224.70d
Venus to revolve around the Sun

225d
A day on Venus—*compared to an Earth day*

The growing season in Washington, D.C.

226d
The gestation of the chimpanzee

227d
The clay-colored sparrow to winter in Sonora

TWO HUNDRED THIRTY-ONE TO TWO HUNDRED THIRTY-NINE DAYS

231d
The growing season in Philadelphia, Pennsylvania

233d
The growing season in Atlanta, Georgia

234d
The growing season in Jackson, Mississippi

236d
The growing season in Memphis, Tennessee

237d
The rocket flight from Earth to Mars

238d
The growing season in Charlotte, North Carolina, and Norfolk, Virginia

TWO HUNDRED FORTY DAYS

Maximum extension given individuals for filing late on their income tax—*Sixty days automatic extensions are given after the April 15th deadline by filing form 4968 with an estimated tax and two to six months additional time may be granted with due reason.*

Sheep and dogs to reach sexual maturity

Period that venison will retain its freshness in the freezer

The maximum time a turkey should be kept in the freezer

The gestation of the pine marten

Homemade beer to age so that it is potable if malt is made at 143° F.

For the human body to reach the limit of its tolerance to 3 roentgens of protracted radiation exposure a day

EIGHT TO NINE MONTHS

A baby to learn that dropping things is fun—*especially when someone else picks them up for him!*

TWO HUNDRED FORTY-ONE TO
TWO HUNDRED FORTY-NINE DAYS

241d
The siege of Port Arthur by the Japanese—*The Russian garrison surrendered January 2, 1905.*

242d
The growing season in El Paso, Texas

243d
Venus to make one rotation (retrograde) on its axis

245d
Young gorillas to begin walking

TWO HUNDRED FIFTY DAYS

6,000h
Life of the fluorescent tube used in Plug Tote-A-Lite —*A portable light which will operate on a 12-volt lighter or battery pack.*

250.59d
Jupiter's sixth satellite to make one revolution

TWO HUNDRED FIFTY-ONE TO
TWO HUNDRED FIFTY-NINE DAYS

255d
Jupiter's tenth satellite to make one revolution

258d
The gestation of the yak

259d
The human fetus to be considered full term—*less than 37 weeks is considered premature*

259.7d
Jupiter's seventh satellite to make one revolution

TWO HUNDRED SIXTY-ONE TO
TWO HUNDRED SIXTY-NINE DAYS

267d
The 1967–68 Detroit newspaper strike

TWO HUNDRED SEVENTY DAYS

Period between graduation and first payment due date on a national direct student loan (see also 10y)

The gestation of the dolphin

The annual term of the U.S. Supreme Court hearings —*During that time they hear oral arguments of one hour's duration, four hours a day, three days a week, two weeks a month in nearly 200 cases a term.*

The breeding season of the female cotton rat in southern latitudes (see also 7mo and 12mo)

To age homemade apple cider

To ripen blue cheese to give it a pronounced flavor (see also 3–4mo)

The gestation of the American buffalo

TWO HUNDRED SEVENTY-ONE TO TWO HUNDRED SEVENTY-NINE DAYS

278d
The growing season in Portland, Oregon

TWO HUNDRED EIGHTY DAYS

The growing season in Seattle, Washington

The gestation of the cow

TWO HUNDRED NINETY-ONE TO TWO HUNDRED NINETY-NINE DAYS

291d
The growing season in New Orleans, Louisiana

298d
The growing season in Mobile, Alabama

THREE HUNDRED DAYS

A year according to the Babylonian calendar

To recover from a fracture to the shaft of the thigh-bone

To tow an iceberg from Antarctica to the Atacama desert in Chile—*As world sources of fresh water dwindle, the venture may prove profitable even though the berg may lose 86% of its volume. The 35 million cubic meters remaining at the end of the voyage would be worth 2.7 million dollars.*

Average prison term for a person convicted of liquor laws violations

Life of semen in female guppies after mating

The average marriage engagement

To age Parmesan cheese—*FDA regulation*

Widows of the Marshall Bennett Islands to prepare for ritual purification at the seashore by their sisters-in-law

Homemade beer to age so that it is potable if malt is made at 148° F.

To prepare the federal budget for any one fiscal year

THREE HUNDRED NINE DAYS

The growing season in Houston, Texas

THREE HUNDRED FOURTEEN DAYS

The growing season in Jacksonville, Florida

THREE HUNDRED TWENTY DAYS

The growing season in Sacramento, California

THREE HUNDRED THIRTY DAYS

NASA's unmanned spaceflight of Project Viking to make a flight from Cape Canaveral to an orbit around Mars—*1975–1976*

An infant to begin walking when led

The gestation of the alpaca and the horse

The incubation of the king crab

ELEVEN TO TWELVE MONTHS

The gestation of the zebra ·

FIFTY WEEKS

Playing season of the fully endowed Boston Symphony Orchestra (see also 26w)

THREE HUNDRED FIFTY-THREE TO THREE HUNDRED EIGHTY-FIVE DAYS

A year on the Hebrew calendar

THREE HUNDRED FIFTY-FIVE DAYS

A Mohammedan year when an intercalary day is added—*eleven out of every thirty Mohammedan years*

A year on the Numa calendar

THREE HUNDRED SIXTY DAYS

The gestation of the fin whale

The satellite Nereid to make one revolution around Neptune

HOW LONG DOES ONE SOLAR YEAR TAKE?

365d, 5h, 48m, 45.51s

ONE YEAR

365d, 5h, 47m
The Mayan calendar year

365d, 5h, 48m
The astronomical year

365d, 5h, 48m, 46.43s
The tropical year—*from winter solstice to winter solstice*

365, 5h, 49m, 12s
The Gregorian calendar year

365d, 6h
The Julian calendar year

Maximum incubation period of rabies (see also 10d, 30d, 50d)—*30–50 days is average*

To raise lamb for slaughter

Ten to twenty million people to die of starvation

A dripping water faucet to waste 900 gallons of water
—*dripping at the rate of drop per second*

Rabbit incisors to grow 20 inches—*10 feet of growth
during its lifetime*

The south Atlantic Ocean to widen between South
America and Africa on an average of 2 inches

A pack-a-day smoker to smoke 7,300 cigarettes

Minimum expected illness that qualifies a U.S. worker
to apply for disability from Social Security

Los Angeles to move 2 to 3 inches toward San Francisco—*Los Angeles is on the west side of the San
Andreas fault and San Francisco is on the east side*

The human heart to pump 730,000 gallons of blood

Period during which female cotton rats in captivity
will breed (see also 7mo and 9mo)—*compared with
rats in the wild*

A hen to lay an average of 220 eggs

An average of 25,000 newly patented inventions to be
developed in the United States

Infants to reach age when they are usually given their
combined measles-mumps-rubella shot and first tuberculin skin test

Young ground squirrels to get the mating urge—
*although they do not reach maximum physical size
for 2 years*

The period Congo natives wait to sweep out their
houses after there has been a death in the family—

so the sweeping will not injure the ghost of the departed

U.S. taxpayers to lose $428 million in interest because the federal government keeps their money—*$3.9 billion in noninterest paying bank accounts*

The force of gravity to grow .1 billionth part weaker as the orbit of the Moon grows gradually longer

A fiscal year

The bovine population of the United States to burp 50 million tons of hydrocarbons into the atmosphere

A gopher's teeth to grow 40 inches in length

A blood donor to safely give 5 pints of whole blood or 26 pints of plasma

The growing season in Honolulu, Hawaii; Los Angeles and San Francisco, California; and Miami, Florida

The average public defender to handle 150 felony cases, 400 misdemeanors, 200 juvenile, 200 cases under the mental health act, and 25 appeals

To build a Steinway grand piano

The Egyptian year to fall behind the sun by 6 hours

The per capita consumption of 3 pounds of tuna in the United States

A mouse to become a senior citizen

The Gregorian calendar to gain 25 seconds on the sun

The average American home to burn 30 to 60 cords of wood when fireplaces were used exclusively for heating and cooking

One hundred billion tons of carbon to undergo photosynthesis worldwide

A family of eagles to consume 550 pounds of flesh— *parents and two young*

The guarantee on parts and labor on solid-state color televisions

A cocoa tree bearing six thousand flowers in a single season to produce a mere twenty to forty pods containing harvestable cocoa beans

One half million people to visit the Empire State Building

Dietary restriction to end following gall bladder surgery

350 million toy marbles to be produced in the United States

The world's largest refuse incinerator to process 700 thousand tons of refuse—*located in Rotterdam, The Netherlands*

Seventy percent of patients with diagnosed acute leukemia (see also 3y) and 59 percent with chronic leukemia to die

An American child to consume 3 to 6 pounds of salt

Man to remove the following minerals from the earth: petroleum, 600 million tons; coal, 2 billion tons; iron, 200 million tons; lead, 4 million tons

An 18-month-old child to increase vocabulary from 22 to 446 words and a child between 7 and 14 years old to increase vocabulary by 700 words—*M. S. Smith study at the University of Iowa*

Caucasian babies to catch up to Negro babies in motor development—*tests have shown black infants display motor superiority during the first year of life*

Between January 1, 1 B.C. and January 1, 1 A.D.—there *was no year zero*

For 42.7% of persons released after serving a prison term to be rearrested

Effective time of the rabies and distemper shots for household pets

Light to travel 5 trillion miles—*one light year*

The vernal equinox to change 50.3 seconds

Thighbone (femur) fractures to heal—*either neck or shaft of the bone*

16 million thunderstorms to occur on Earth

Doctors in the United States to write over 2 billion prescriptions

900,000 abortions to be performed in the United States —*making it the second most frequent surgery after tonsillectomy*

The national cash surplus in the United States to reach over $10 billion if every American saved $1 a week—*ad campaign of Equitable Trust Company*

To forget what you learned in school—*based on a classic study by Herman Ebbinghaus: one year after the end of each course to forget 79% Botany; 61% Zoology; 76% Psychology; 44% Algebra; 40% Latin; 53% Chemistry, and 44% History*

Allowable filing period for federal estate tax following the decedent's date of death

Period of effective immunization against influenza

Maximum safe freezer storage for fruits and vegetables

Period between moves for 17.9% of Americans

Warranty of most GE appliances

A quarter of a million foreign cars to be unloaded at the port of Baltimore—*number one port of entry for automobiles in U.S.*

One hundred laying hens to produce 2,400 pounds of manure—*or 4.51 pounds per bird*

A Virginia oyster to reach sexual maturity

Validity of an airline ticket after issuance

Fashionable clothes to be seen as "dowdy"—*according to James Laver's Time Spirit Theory (see also 10, 20, 30, 50, 70, 100 and 150 years)*

A dairy cow to produce milk (in kilograms): Mexico, 2,347; Netherlands, 4,246; United States, 4,038; U.S.S.R., 2,085; Argentina, 460; United Arab Republic, 675; Spain, 1,610

Eleven to twelve million surgical operations to be performed in the United States

Pregnancy to be advisable following removal of a kidney stone

Parsnip wine to age in the cask

Minimum time between local influenza epidemics (see also 4y)

Children to outgrow these toys (listed with ages at which they appeal): mobile—one month to one year; teething rings—six months to one year; push and pull toys—one year to two years; 25-piece jigsaw puzzles— three years to four years—*prepared by* Where, *British consumer magazine*

52w, 1d
The calendar first based on a seven-day week—*perfected about 1500* B.C. *and attributed to Moses*

1y, 2d
The synodic period of Neptune and Pluto

1y, 5d
The synodic period of Uranus

1y, 13d
The synodic period of Saturn

THIRTEEN MONTHS

A child to begin to crawl up and down stairs

Average prison term for violation of the Selective Service Act

1y, 34d
The synodic period of Jupiter

FOURTEEN MONTHS

The siege of La Rochelle by the forces of Richelieu
—*the Huguenots surrendered October 28, 1628*

The period homesteaders must reside on claimed land
in order to gain title from the U.S. government—
*Some payment is involved in this minimum period
whereas no payment is involved if homesteading is
continued for five consecutive years.*

A marmoset, harbor porpoise and squirrel monkey to
reach sexual maturity

A child to begin standing alone, unaided

Minimum time to ripen Italian Parmesan cheese

FIFTEEN MONTHS

To age homemade beer so that it is drinkable—*if malt
is made at 152° F.*

Interval between dental visits among children of
lower-income families (see also 155d)

The United States to be served by three vice-presidents
—*Spiro Agnew, Gerald Ford, Nelson Rockefeller*

A tadpole to grow into a bullfrog

15mo, 3d
An infant's lower first molar to erupt

15mo, 8d, 8h
The amount of classroom time in hours spent in high
school—*11,000 hours*

SIXTEEN MONTHS

Period of potency for semen stored in a sperm bank

16mo, 18d
The full course of a criminal appeal in California—
1970 study by Winslow Christian

SEVENTEEN MONTHS

Cattle to reach prime slaughter age

EIGHTEEN MONTHS

Ninety percent of women in their late teens and early twenties to get pregnant when they have intercourse every other day

To reach minimum ideal age to surgically correct a cleft palate (see also 2½y)

To cure tubercular pleurisy

Infants to reach the age when they are usually given their first booster shot for diphtheria-tetanus-pertussis, and their first oral polio vaccine booster

A one-dollar bill to wear out

The minimum growing life of a strand of human hair—*some are said to grow as long as 6 years*

The human infant's sensory period according to Piaget's series of stages

Schooling for an aircraft mechanic

The maximum construction time for a home in which profits from a former home are invested in order to qualify for deferment of capital gains tax

Minimum operable age for congenital clubfoot if non-operative treatment fails (see also 3y)

Children to reach age when corrective eye surgery is feasible

Minimum accrued work time to be fully insured by Social Security (see also 10y)

Rejuvenating effects of chemical skin peeling to wear off

Phenothiazine derivatives to be excreted from the body —*drug used to tame destructive and assaultive psychotics*

To market flax after harvest

Children to outgrow these toys (listed with ages at which they appeal): rattles—four months to 18 months; crib and carriage beads—four months to two years; Grippable ball—six months to two years; unbreakable mirror—nine months to two years; screwing toys—18 months to three years; hammer pegs—18 months to three years; low riding or rocking toys—18 months to three years—*prepared by* Where, *British consumer magazine*

584d
The synodic period of Venus

NINETEEN MONTHS

The average prison term served by counterfeiters and forgers

Suckling period for Zulu mothers

TWENTY MONTHS

15,100h
The time in hours that teenagers spend watching television during their high-school years

TWENTY-ONE MONTHS

The gestation of the elephant

A child to lose most of his "baby fat"—*from a peak of fat at 9 months to 2½ years*

1y, 266d
Jupiter's twelfth satellite to make one revolution

TWENTY-TWO MONTHS

1y, 10mo, 21d
Mars to revolve around the Sun

TWO YEARS

Maximum time to have a U.S. patent granted (see also 6mo)

Paralysis to leave muscles following an attack of poliomyelitis—*maximum time*

Time between shearings of the South American alpaca

To age three-star Cognacs

Forty-five scriveners to hand copy 200 volumes—*before the Gutenberg press*

U.S. representative's congressional term—*commencing on January 3d of each odd-numbered year*

Parts guarantee on a Maytag dishwasher

For U.S. manufacturers to meet the demand for nylon stockings after the end of World War II

Interval between cleanings of septic tanks operating at normal family usage

Suckling period for elephants

To reach the cotton wedding anniversary

To tabulate the U.S. census in 1890 (see also 7y)—*a total of 63,056,000 people, using Herman Hollerith's punch card tabulator*

A governor's term in the following states: Arizona, Arkansas, Iowa, Kansas, New Hampshire, New Mexico, Rhode Island, South Dakota, Texas, Vermont and Wisconsin (see also 4y)

52.5% of persons released after serving a prison term to be rearrested

Period of human pubescence

Average period for performing corrective orthodontia

Driving experience necessary to become a local transit bus driver

To cure most types of tuberculosis with chemotherapy

Shelf life of cyalume chemical light sticks

Children to reach the age when they can first wear eyeglasses

To become an embalmer—*college-level work*

A foreign child adopted by U.S. citizens to become naturalized

Guarantee on picture tubes for most solid-state color televisions

A blackberry vine to grow into an uncontrollable thicket

Period a new color television should give trouble-free service

Validity on drivers' licenses in the following states: Alabama, Arkansas, Connecticut, Florida, Georgia, Iowa, Kentucky, Louisiana, Maine, Maryland, Mississippi, New Mexico, North Dakota, Oklahoma, Oregon, Pennsylvania, Rhode Island, Tennessee, Vermont, Washington, and Wisconsin

Young lions to become independent

Cheddar cheese to reach its flavor peak

A child to develop his own color preferences—*study by John Watson*

A child to reach about half his adult height (birth to 2 years old)

A child's preference for left or right handedness to become well established

To complete a lather's apprenticeship

Minimum period during a young man's life when he is likely to engage in sexual acts with animals (see also 3y)—*Alfred C. Kinsey reports instances of such behavior continuing for as long as 10 to 15 years, or even for life*

The earliest a child begins speaking in sentences and complying with demands (see also 3y)

Minimum time for testosterone therapy to cure testicular failure in males when puberty does not occur normally (see also 3y)

Suckling period for walruses

Rotation of tobacco crops

To become a railroad signalman

Children to outgrow these toys (listed with ages at which they appeal): baby walker—one to three years; teddy bear—one to three years; hobby horse—three to five years; 50-piece jigsaw puzzles—four to six years; doctor and nurse kits—four to six years—*prepared by* Where, *British consumer magazine*

TWO TO THREE YEARS

2y, 28d
Jupiter's ninth satellite to make one revolution

2y, 50d
The synodic period of Mars

2y, 2mo
An infant to acquire his lower second molar

2y, 6mo
An adolescent girl to grow an average of 3¼ inches
and an adolescent boy to grow 8 inches—*during peak
growth periods*

The human birth weight to quadruple

Maximum tour of duty by Chinese laundrymen aboard
an English ship

The Bordeaux region of France to produce clarets in
the cask (see also 2y)

Tequila to ferment

2y, 329d
The half-life of the radioisotope iron 55 (see also 46d,
7h, 12m)

2y, 351d
The Magellan expedition to make the first trip around
the world—*September 20, 1519 to September 6, 1522;
Ferdinand Magellan died en route.*

THREE YEARS

To age Scotch whiskey—*minimum set by the Imma-
ture Spirits Act of Great Britain*

The incubation of leprosy

Enroll period for Medicare medical insurance once
eligibility begins

To complete a roofer's apprenticeship

The period that 31.8% of the U.S. population remains
in a single residence before moving

An asparagus plant to reach harvestable age—*then only for a period of 3 to 5 weeks. Mature plants may be cut up to 8 weeks a season*

58.3% of persons released after serving a prison term to be rearrested

Young buffalo bulls to get the mating urge—*although they do not reach full physical maturity until 8 years old*

The accuracy of a quartz crystal clock to vary 1 second

To pay off the final quarter of a 30-year, self-liquidating mortgage at 10% interest; the first quarter of a 10-year mortgage at 8%; the final quarter of a 20-year mortgage at 8% and the final quarter of a 30-year mortgage also at 8% interest

To reach the leather wedding anniversary

To earn a flight engineer's license

The economic life expectancy of dump truck tires—*run 2,000 hours*

For the human brain to reach 80% of its adult size

The validity of a broadcast license granted by the Federal Communications Commission

To repay a standard auto loan—*some now as long as 48 months as prices increase*

For a child to develop mentally to the extent that he can identify emotions expressed in facial expressions, e.g., laughter—*study by Landis*

A girl of the Punan tribe of Borneo to reach the age when she celebrates her maturity, when she is weaned, and begins smoking a pipe and wearing earrings

To complete an apprenticeship as an automobile painter, stone mason, or iron worker

A concrete finisher's apprenticeship

Average terminal stay in a U.S. nursing home—*study by Mary Mendelson, author of* Tender Loving Greed *and Cleveland community planning consultant*

A painter or paper hanger's apprenticeship

A construction machinery operator's apprenticeship

A glazier's apprenticeship

A young sulphur-bottom whale to grow to a length of 80 feet

Warranty on Lady Sunbeam shavers

Maximum period during a young man's life during which he engages in sexual acts with animals (see also 2y)—*Alfred C. Kinsey reports instances of such behavior continuing for as long as 10 to 15 years, or even for life*

A lobster claw to regenerate

The effective period of redwood stain when exposed constantly to the elements (see also 24h)—*developed by the forest products laboratory of the U.S. Department of Agriculture. The durable stain combines paraffin, zinc stearate, turpentine, pentachlorophenol, boiled linseed oil, and pigment.*

The fin whale to reach sexual maturity

Men to die from specified types of cancer following diagnosis: stomach, 86%; colon, 51%; rectum, 55%; lung and bronchus, 90%; prostate, 36%; kidney, 57%; bladder, 39%; skin, 37%—*data compiled by the Metropolitan Life Insurance Company with the aid of End Results Evaluation Program of the National Cancer Institute*

Period of validity on drivers' licenses in the following states: Alaska, Arizona, Colorado, Idaho, Illinois, Michigan, Missouri, New Jersey, Ohio, and Wyoming

To become a first lieutenant in the U.S. Army

Minimum period between local chicken pox epidemics

Spouse of a U.S. citizen to become naturalized

Women to die from specified types of cancer following diagnosis: breast, 27%; stomach, 83%; colon, 49%; rectum, 50%; lung and bronchus, 85%; cervical, 36%; uterus, 24%; kidney, 55%; bladder, 27%; skin, 21%

A race horse to reach its prime running age

Children to outgrow these toys (listed with ages at which they appeal): foam ball—six months to three years; nesting blocks—nine months to three years; graded threading rings—18 months to four years; plastic or wooden train set—three years to six years; finger or glove puppets—three years to six years; farm machinery—four years to seven years; sewing machine —seven years to ten years—*prepared by* Where, *British consumer magazine*

THREE TO FOUR YEARS

3.3y
The comet Encke to orbit the Sun

3y, 6mo
To properly make and age French vermouth

3.6y
Japanese manufacturers to produce and market a new invention—*see also 5.6, 6.4 and 7.3 years*

3.63y
The asteroid Vestat to revolve around the Sun

Maximum time for testosterone therapy to cure testicular failure in young males when puberty does not occur normally (see also 2y)

THREE TO FIVE YEARS

Period that breast cancers remain undetected in most women—*according to Dr. Benjamin F. Byrd, Jr., president of the American Cancer Society*

FOUR YEARS

A musk deer cow to reach sexual maturity

A child's birth height to double

The 29th of February to recur—*Gregorian calendar, Leap Day*

Children to reach average weight and height for their nationalities: Chinese, 14.9 kg., 98 cm.; Indians, 12.4 kg., 91.6 cm.; African Negroes, 12.2 kg., 91.6 cm.; Czechs, 16.3 kg., 101.5 cm.

To complete any of these trade apprenticeships: mill-wright, forger, boiler maker, tool and die maker, machinist, sheet metal worker, asbestos and insulation worker, carpenter, and electrician

To become a trainer of seeing-eye dogs

The area around Kawasaki, Japan, to rise two inches —*the bulge, experts say, is due to the pressure from the earth prior to an earthquake*

The life of pine needles

Children to outgrow these toys (listed with ages at which they appeal):rag doll—one to five years; telephone—two to six years; baby doll—two to six years; big crayons—2½ to six years; wooden constructional sets—three to seven years; stuffed soft toys—three to seven years; childsize carpentry tools—six to ten years; chemistry set—ten to 14 years; plastic model kits (1/24 scale and above)—11 to 15 years—*prepared by* Where, *British consumer magazine*

To produce gold tequila which is shipped to the United States still in the vats—*white tequila is not aged*

The period of validity on drivers' licenses in the following states: California, Delaware, Hawaii, Indiana, Iowa, Kansas, Massachusetts, Minnesota, Montana, Nebraska, Nevada, New Hampshire, North Carolina, South Carolina, South Dakota, Texas, Utah, Virginia, West Virginia, and the District of Columbia

Macaques to reach sexual maturity

Human adolescent males to grow pubic hair

The interval between Olympiads

An oyster to reach commercial size

To age V.S.O.P. Cognac—*Very Special Old Pale*

To arrest an advanced case of leprosy with drug therapy

Dwarf plum trees to begin bearing fruit

To reach the fruit and flower wedding anniversary

To become a teacher—*minimum college*

To become an insurance underwriter—*college degree*

A quadrennium

To pay off these portions of self-liquidating mortgages at 8% interest: the second half of a 10-year term; the third fourth of a 20-year term; and these portions at 10% interest: the third and second fourths of a 20-year term; the third fourth of a 30-year term

Maximum time between local influenza epidemics (see also 1y)

The dwarf cherry tree to begin yielding fruit

To age Scotch whiskey imported to the United States without age labels (see also 3y)—*minimum U.S. government standard*

FOUR TO FIVE YEARS

4.3y
Light to travel from the nearest star to Earth

4.36y
The asteroid Juno to revolve around the Sun

4.5y
The period during which Auschwitz, the Nazi crematorium, disposed of 2,200 Jews a day

4.5y
Average prison term served for robbery

4.6y
The asteroid Ceres to revolve around the Sun

Children to reach the age when they should have their second booster shot for diphtheria-tetanus-pertussis and a booster dose of oral polio vaccine

The dogwood tree to begin flowering

Children born with Tay-Sachs disease to die

The female baboon to reach sexual maturity

FIVE YEARS

The guarantee on the sealed refrigeration system of most dehumidifiers

The life expectancy of a bulldozer—*driven 2,000 hours a year*

The maximum prison term for evasion of income tax

Children to outgrow these toys (listed with ages at which they appeal): music box—18 months to six years; small diecast vehicles—18 months to six years; wheelbarrow—three to eight years; easel/blackboard —three to eight years; farm and animals—three to eight years; doll house and furniture—five to ten years; modeling clay—seven to 12 years—*prepared by* Where, *British consumer magazine*

A dwarf peach tree to reach its maximum fruit yield

The maximum prison term for importing or mailing pornography

An immigrant to the United States to prove his loyalty for purposes of becoming a naturalized citizen

The military commitment of a graduate from West Point or the U.S. Naval Academy

The guarantee on cabinet rusting in Maytag dish-washers, the motor of Kitchen Aid dishwashers, and some major components of the Waste King universal dishwasher

Mozart to become proficient at piano, organ, violin, and musical composition—*between the ages of three and eight*

The period homesteaders must reside on claimed property in order to qualify for title of the property without payment

The number of years a home must be occupied by the owner out of 8 years preceding its sale if the owner (having reached his 65th birthday) wishes to qualify for his once-in-a-lifetime privilege of escaping capital gains attributed to the first $35 thousand of the sale price

The validity of a U.S. passport

To gain a marriage annulment in New York after either party is declared incurably insane

The Roman census period—*lustrum*

To complete a photoengraver's apprenticeship

To complete a plumber and pipe fitter's apprenticeship

A voter's registration to run out if the option to vote is not exercised

To build the Verrazano Narrows bridge in New York (see also 6y)—*one of the world's longest and highest suspension bridges*

Women to die from specified types of cancer following diagnosis: breast, 37%; colon, 54%; rectum, 59%; cervix, 43%; uterus, 28%; kidney, 62%; bladder, 30%; skin, 27%—*data compiled by the Metropolitan Life Insurance Company with the aid of End Results Evaluation Program of the National Cancer Institute*

Men to die from specified types of cancer following diagnosis: stomach, 90%; colon, 58%; rectum, 63%; lung and bronchus, 92%; prostate, 48%; kidney, 64%; bladder, 44%; skin, 45%—*data compiled by the Metropolitan Life Insurance Company with the aid of End Results Evaluation Program of the National Cancer Institute*

The interval between floodings of Lake Kurnas on Crete

The International Paper Company to plant a tree for every American

Children to develop the ability to distinguish left from right relating to objects and other people—*between the ages of five and ten*

The minimum work time during the 10 years preceding an illness or injury which qualifies a U.S. worker to collect disability from Social Security

The Common Market to give $4.1 billion in financial aid to forty-six developing countries in return for trade concessions—*1975–1980*

A rubber tree to grow large enough to tap

Drug arrests among women to increase over 1,000% —*1969–1973*

A monthly deposit of $25 to total $1,700 and a monthly deposit of $50 to total $3,400 in a savings account—*based on 5¼% interest compounded daily*

50% of boys to reach age when they are able to achieve sexual climax

The American holly tree to begin flowering

The maximum life span of plants in a terrarium

5y, 109h, 12m
Half-life of the radioisotope cobalt 60

5.6y
West German manufacturers to produce and market a new invention—*also see 3.9, 6.4 and 7.3 years*

SIX YEARS

Life span of snails—*and most other gastropods*

A musk deer bull to reach sexual maturity

Most cases of muscular dystrophy to be discovered after birth

The life span of the cardinal

To pay off the second quarter of a 30-year self-liquidating mortgage at 10% interest and the first half of a 10-year mortgage at 8% interest

Children to reach the age when they are first advised to use adult car seat belts

Term of state supreme court judges in the following states: Alabama, Arizona, Florida, Georgia, Idaho, Indiana, Kansas, Minnesota, Montana, Nebraska, Nevada, Ohio, Oklahoma, Oregon, South Dakota, Texas, and Washington

A child's visual acuity to approach 20/20 after birth —*corresponds roughly to the maturation of the macula of the retina*

Term of a U.S. senator

A dwarf apple tree to begin bearing fruit

A female gorilla to reach sexual maturity

The human brain to reach 90% of its adult weight

Children to develop binocular vision

Maximum time to make a geological study of an oil field

To ripen Chianti in vats to the point where it is mellow and soft

The statute of limitations on federal income tax returns

Children to outgrow these toys (listed with ages at which they appeal): poster paints—2½ to eight years; wooden blocks—three to nine years; nice dressable doll—six to 12 years; board games with tactical skill —eight to 14 years; powered models, motor racing, trains, planes—nine to 15 years—*prepared by* Where, *British consumer magazine*

Children to mature enough to grasp number concepts —*not just recite by rote*

A child's heart rate to approach that of a normal adult—*70 to 100 beats per minute*

6y, 2mo
Waiting period for public housing in the District of Columbia

6.4y
United States manufacturers to produce and market a new invention—*also see 3.6, 5.6 and 7.3 years*

6.5y
The comet Giacobini Zinner to orbit the Sun

Increased life expectancy of a 25-year-old who is a nonsmoker compared to one who smokes a pack of cigarettes a day

6y, 10mo
A kidnapper to serve a prison term

SEVEN YEARS

To become a captain in the U.S. Navy

Bad luck curse after breaking a mirror

A Hoolock gibbon to reach sexual maturity

To reach the wool and copper wedding anniversary

Term of office of a Federal Communications Commissioner—*appointed by the President*

The maximum agreed upon time for ratifying a Constitutional amendment—*In recent years Congress has interpreted the "reasonable time after proposal" stipulated by the Supreme Court to be this period.*

Every cell in the human body to be renewed

Information regarding an individual's credit rating to be removed from record (see also 14y)—*stipulation of the Fair Credit Reporting Act does not include bankruptcy*

A missing person to be declared legally dead

Between sabbatical years—*Hebrew tradition*

Maximum term of a loan to farmers for operating expenses from the Farmers Home Administration—*U.S. Department of Agriculture*

Period of citizenship required of those running for office as a U.S. representative

Former smoker Mrs. Clair Manjack of Grand Rapids, Michigan, to accumulate $1,500 by saving the cost of a carton of cigarettes every week

To pay off the last half of a 30-year self-liquidating mortgage at 10% interest and the second quarter of an equal term mortgage at 8%

A $3,500 investment in a solar energy unit and backup system to pay for itself in fuel savings—*at present rates*

A girl to reach 75% of her adult height (see also 9y)

Term of state supreme court judges in Hawaii and Maine

A bull elephant seal to gain the experience necessary to protect a harem against other bulls

To tabulate the U.S. census in 1880 (see also 2y)—*a total of 50,262,000 people*

A person suffering from pituitary dwarfism to grow 14 inches after beginning therapy with growth hormone

King crabs to reach legal size in the Bering Sea— *7 inches*

7.2y
A marriage which ends in divorce in the United States —*median duration*

7.3y
French manufacturers to produce and market a new invention—*see also 3.6, 5.6 and 6.4 years*

7.42y
The comets Whipple and Faye to orbit the Sun

EIGHT YEARS

To become proficient at the slight of hand necessary to execute the classic cups and balls magic trick—*such as the shell game. As Washington area magician Al Cohen puts it, "Any six-year-old can do it after eight years practice."*

Nine out of ten men 40 years old with a systolic blood pressure of 165 and blood cholesterol of 285 (both high) to have a stroke or heart attack—*as opposed to four out of 100 men of that age with normal levels*

Interval from the geological survey of an oil field through the stages of oil retrieval and refinement until gasoline can be pumped into an automobile—*experience of Exxon in exploring and drilling for oil in the Gulf of Mexico*

To pay off the first quarter of a 20-year self-liquidating mortgage at 8% interest and the second half of a 30-year note also at 8%

To reach the bronze and pottery wedding anniversary

To complete the compulsory school attendance required by these states: Arizona, California, and Washington

Children to outgrow an outdoor slide or swing—*prepared by Where, British consumer magazine*

Term of state supreme court judges in the following states: Arkansas, Connecticut, Iowa, Kentucky, Michigan, Mississippi, New Mexico, North Carolina, Tennessee, and Wyoming

Children to reach average weight and height for their nationalities: Chinese, 20.8 kg., 119.2 cm.; Indians,

18.7 kg., 116.1 cm.; African Negroes, 21.4 kg., 121.1 cm.; Czechs, 25.2 kg., 125.6 cm.

A child to gain optimal benefits from fluorinated water

8.57y
The comet Sola to orbit the Sun

NINE YEARS

Term of office for members of the board of governors of the U.S. Postal Service—*Presidential appointments*

Period of citizenship required of those running for office as U.S. senators

Compulsory school attendance in the following states: Alaska, Arkansas, Colorado, Connecticut, Florida, Georgia, Illinois, Indiana, Iowa, Kansas, Kentucky, Louisiana, Maryland, Massachusetts, Minnesota, Mississippi, Missouri, Montana, Nebraska, New Mexico, North Carolina, North Dakota, Pennsylvania, Rhode Island, South Dakota, Vermont, West Virginia, Wyoming, and the District of Columbia

A boy to reach 75% of his adult height (see also 7y)

To pay off the first quarter of a 20-year self-liquidating mortgage at 10% interest

Children to attain the age when they can care for their own artificial eyes

TEN YEARS

A decade

2,179,886 boys and girls to win the presidential award for youth fitness—*1965–1975*

Maximum effectiveness of tetanus immunization

A dwarf peach tree to end its fruit-bearing years

Termite protection from soil poisons—*such as aldrin, chlordane, or dieldrin*

Flourocarbons released at the Earth's surface to reach the ozone layer (see also 100y)—*Fluorocarbons released from aerosols or refrigeration units*

Potentially the longest term of a U.S. President—*He may serve up to two years of an uncompleted term and still be elected to two, four-year terms.*

The "cavity-prone years"—*ages five to fifteen*

Metric conversion in the U.S.—*according to a bill signed by President Gerald Ford in 1975*

Effective period of diphtheria immunization—*boosters should continue through adult life*

Maximum prison sentence for passing bad checks

Term of state supreme court judges in Alaska, Colorado, Illinois, North Dakota, South Carolina, Utah, and Wisconsin

Between major U.S. census polls

Compulsory school attendance required by Alabama, Delaware, Maine, Michigan, Nevada, New Hampshire, New Jersey, New York, Tennessee, Virginia, and Texas

Japanese artisan Ryumo Hori to complete one of her minutely carved wooden dolls clothed in handwoven costumes—*Ryumo Hori has been designated one of Japan's living national treasures*

Term for repaying a national direct student loan after graduation

Life span of a mink

Women to die from specified types of cancer following diagnosis: breast, 49%; colon, 60%; rectum, 62%; cervix, 48%; uterus, 33%; kidney, 68%; bladder, 31%; skin, 32%—*data compiled by the Metropolitan Life Insurance Company with the aid of End Results Evaluation Program of the National Cancer Institute*

Average road life of an American automobile

Maximum work time to become "fully insured" under Social Security (see also 18mo)—*forty 3-month quarters*

To recover the world supply of gold, platinum, lead, and zinc—*at present consumption rates*

Minimum time for lung and skin cancer to show up in factory workers exposed to chloroprene—*a colorless liquid chemical used in the production of synthetic rubber. (Recent studies in the Soviet Union)*

A former member of a subversive organization to prove his loyalty to the United States for purposes of becoming a naturalized citizen (see also 5y)

An elk to reach old age

Nature to restore the annual vegetation in those areas of the Mohave Desert used for the 155-mile Barstow to Las Vegas, Hare and Hound motorcycle race—*estimates of damage to perennial trees and shrubs run into the centuries*

Fashionable clothes to be seen as "hideous"—*according to James Laver's Time Spirit Theory, see also 1, 20, 30, 50, 70, 100 and 150 years*

Children to outgrow Leg-o building blocks—four to 14 years—*prepared by* Where, *British consumer magazine*

Men to die from specified types of cancer following diagnosis: stomach, 92%; colon, 63%; rectum, 71%; lung and bronchus, 95%; prostate, 63%; kidney, 73%; bladder, 52%; skin, 52%—*data compiled by the Metropolitan Life Insurance Company with the aid of End Results Evaluation Program of the National Cancer Institute*

The red alder and paper birch trees to begin flowering

Eighty percent of preadolescent boys to attain the age when they are able to reach climax—*Alfred C. Kinsey, University of Indiana study*

Flowering to begin in the following varieties of trees: savin juniper, white spruce, giant, arborvitae, pecan, southern poplar, and black cherry

Peak of the fruit-bearing years in the dwarf cherry and plum trees

Minimum lag time for most forms of cancer (see also 20y)

To become an experienced airline dispatcher

To reach the age when the first symptoms of Wilson's disease appear

10.5y
The comet Vaisala to orbit the Sun

ELEVEN YEARS

Compulsory school attendance in these states: Idaho, Oklahoma, Oregon, South Carolina, and Wisconsin

Children to get their permanent teeth—*upper first bicuspid*

To build the Trans-Siberian Railroad—*5,500 miles*

To reach the steel wedding anniversary

The sunspot cycle—*period of minimum activity occurs on an average of every 11.3 years*

11y, 6mo
Maximum allowable amount of turbine creep at a tension of 3 to 4 tons per square inch—*parts should not change size more than .1%*

11.86y
Jupiter to revolve around the Sun

Average widowhood for American women

TWELVE YEARS

The speech area of the human brain's cortex to become fixed—*before the age of 12 the speech center can be relocated if necessary by surgery*

To age Chivas Regal Scotch whiskey

Young elephants to leave their mothers

Jupiter to return to its same relative position in the sky

To reach the silk and linen wedding anniversary

Compulsory school attendance required by Hawaii, Oklahoma, and Vermont

Juveniles to become old enough to obtain a work permit in these states: California and New Hampshire

Peak yielding period of the dwarf apple tree

Harvestable maturity of the mezcal cactus—*The sap of the cactus "pineapple" is used for making tequila.* (Century Plant, American Aloe)

Term of state supreme court judges in California, Delaware, Missouri, and West Virginia

A bull fur seal to become a harem master

Life span of the aardvark

Initial flowering of the red spruce, linden, and American elm

The human thymus gland to regress—*indicating maturity*

12.2y
Schooling of the average American

THIRTEEN YEARS

European girls to reach menarche (see also 17y)—*sexual maturity*

Number of optimal childbearing years for American women

To pay off the first half of a 20-year self-liquidating mortgage at 8% interest

To reach the lace wedding anniversary

A blue whale to reach maturity

Children to reach their peak speed in response time

13.45y
Adolescent boys to reach the mean age at which pubic hair appears

13y, 9mo
Life span of the raccoon

13y, 10.5mo
Boys to reach the mean age at which orgasm results in first ejaculation

5,083d
Number of strike days per 100 workers for the period of 1969–1974 in Italy

FOURTEEN YEARS

To reach the ivory wedding anniversary

Term of the first patent ever recorded in the United States—*issued to Joseph Jenks in 1646 for sawmill equipment*

Maximum time information relating to bankruptcy can remain in an individual's credit report—*stipulation of the Fair Credit Reporting Act*

Term of state supreme court judges in Louisiana and New York

Juveniles to become old enough to obtain a work permit in the following states: Alabama, Arizona, Arkansas, Colorado, Delaware, Illinois, Indiana, Iowa, Kansas, Minnesota, Missouri, Nebraska, Nevada (males), New Mexico, New York, North Dakota, Oregon, Tennessee, Utah, Vermont, Virginia, Washington (males), Wyoming, and the District of Columbia

Average American child to see 18 thousand violent killings on television

A male gorilla to grow to full size (see also 7 to 8y)

To pay off the first half of a 20-year self-liquidating mortgage at 10% interest

Life span of Rin Tin Tin

To become a major in the U.S. Army

The human eye to reach adult size

Children to reach age when they should have their third booster shot for diphtheria-tetanus-pertussis

14y, 5mo
Life span of an otter

14.44y
Boys to reach average age when their voices change

14.49y
Adolescent boys to reach age when they experience a sudden spurt of growth

14y, 9mo
Between Earth sightings of Saturn's rings edge-on

FIFTEEN YEARS

Life span of a queen ant

To reach the crystal wedding anniversary

The world to run out of silver—*based on present estimated recoverable reserves at present mining and consumption rates*

To pay off the first quarter of a 30-year self-liquidating mortgage at 8% interest

Term of a state supreme court judge in Maryland

Ninety-two percent of males and 25% of females to reach the age when they first experience orgasm

Life span of trout (see also 75y)

A nation to go to war after analysis of current literature shows a greater need for power than love—*The theory belongs to Harvard psychology professor David C. McClelland, as reported in the January 1975 issue of* Psychology Today.

The number of reported cases of measles (rubella) in the United States to decrease 96.5%—*after measles vaccine was introduced in 1963*

Dwarf cherry and plum trees to reach the end of their fruit-bearing years

Sixty to seventy percent of Hodgkin's disease patients to die when x-ray treatment is used

Minimum number of years an asparagus plant will yield (see also 20y)

Life span of a leopard—*in the wild*

Flowering to begin in the sitka spruce and longleaf pine

Minimum time for the giant sequoia tree to begin flowering (see also 50y)

15y, 6mo
Life span of the pine marten

Life span of the Rocky Mountain sheep

15y, 8mo
Life span of the chamois

Life span of the gazelle

SIXTEEN YEARS

Age span between the youngest and oldest cases of first ejaculation among young males

Mattel toys to sell one million Barbie Dolls and two million outfits—*introduced in 1958*

Juveniles charged with criminal acts to become old enough to be tried as adults in these states: Alabama, Arkansas, Connecticut, and North Carolina

Juveniles to reach the age when they may obtain a work permit in these states: Connecticut, Florida, Georgia, Hawaii, Louisiana, Maine, Maryland, Massachusetts, Michigan, Montana, Nevada (females), New Jersey, North Carolina, Ohio, Pennsylvania, Rhode Island, Washington (females), and West Virginia

Life span of the cheetah

16y, 10mo
Life span of the antelope

SEVENTEEN YEARS

Life cycle of the cicada (17-year locust)—*longest cycle of known insects*

The average American marriage

European girls to reach menarche during the early 19th century (see also 13y)

Juveniles charged with criminal acts to become old enough to be tried as adults in these states: Colorado, Florida, Georgia, Illinois, Louisiana, Maine, Massachusetts, Mississippi, Missouri, South Carolina, and Texas (males)

To string telegraph lines across the continental United States—*completed in 1861*

To pay off the first quarter of a 30-year self-liquidating mortgage at 10% interest

Children to complete their set of permanent teeth: upper and lower molars

17.49y
Boys to reach their adult height

Life span of the white-tailed deer

17.9y
The comet Neujminl to orbit the Sun

EIGHTEEN YEARS

Ideal age for nasalplasty—*a nose job*

Juveniles to become old enough to obtain a work permit in Wisconsin and Kentucky

To reach the age of legal majority in the following states: Arizona, California, Delaware, Idaho (females), Illinois, Kansas, Kentucky, Maine, Michigan, Montana (females), Nevada (females), New Jersey, Oklahoma, Pennsylvania, South Dakota, Utah (females), Vermont, Virginia, Washington, West Virginia, and Wisconsin

Life span of the king crab

Life span of a goat

To become a Jesuit Priest

Life span of the European genet

Juveniles charged with criminal acts tried as adults in the following states: Alaska, Arizona, Arkansas, Delaware, Hawaii, Idaho, Indiana (traffic violations only), Kansas, Kentucky, Maryland, Michigan, Minnesota, Montana, Nebraska, Nevada, New Hampshire, New Jersey, New Mexico, North Dakota, Ohio, Oklahoma, Oregon, Pennsylvania, Rhode Island, South Dakota, Tennessee, Texas (females), Utah, Vermont, Virginia, Washington, West Virginia, Wisconsin, and the District of Columbia

NINETEEN YEARS

To reach the age of legal majority in Alaska and Iowa

The Hebrew calendar to lag behind the Sun $\frac{3}{4}$ of a day

To attain the age at which delinquency codes are void and those charged with breaking the law are tried as adults in Iowa and New York

To complete the metonic cycle

A puma to live its maximum life span

The nutation period of the Earth

19y, 6 m
The European wild boar to live its maximum life span

TWENTY YEARS

A self-employed individual who contributes the maximum allowable by the federal government to the Keogh retirement plan ($7,500) to accumulate over a quarter of a million dollars—*at 5% interest ($263,240)*

The world-wide need for fresh water and food to double—*according to oceanographers R. H. Charlier and M. Vigneaux*

Life span of a queen bee

Jupiter to overtake Cronos in the sky

Life span of a dog

To reach the china wedding anniversary

The duration of the preservative qualities of pentachlorophenol on fence posts

To gain an extra 60 pounds by eating one slice of bread a day over and beyond the daily calorie intake needed

To collect a million-dollar win on the Maryland lottery—*$50,000 annually*

Maximum number of years an asparagus plant will yield (see also 15y)

The safe shelf life of dehydrated foods

Maximum lag time for most forms of cancer (see also 10y)

Intervals between elections of U.S. Presidents who have died in office beginning with William Henry Harrison in 1840—*When numerologist Walter Gibson proposed this in the 1920s it held true through President Harding. Since that time President Roosevelt—elected to his last complete term in 1940 and President Kennedy in 1960—have followed the pattern which has held true for 120 years. If it continues, the man elected in 1980 will die in office. (Harrison elected 1840; Lincoln, 1860; Garfield, 1880; McKinley, 1900; Harding, 1920; Roosevelt, 1940; Kennedy, 1960.) Notice also that not only is the interval 20 years, but all the dates are divisible by 20.*

Vintage port wine to reach its prime

Per capita consumption of milk to increase 20-fold since the end of World War II in Japan

Life span of sheep

The population of Denver to soar from 4,700 in 1870 to 106,000—*contributing events were the discovery of gold and silver in the region, the routing of remaining Indians, and the arrival of the railroad*

The human body to synthesize 500 to 1,000 pounds of protein—*birth to 20 years old*

A $10 monthly deposit to accumulate $4,200 and a monthly deposit of $100 to accumulate $42,500 in a savings account—*based on 5¼% interest compounded daily*

To develop the transistor—*finally patented in 1948*

To reach the age of legal majority in Hawaii and Nebraska

Fashionable clothes to be seen as "ridiculous"—*according to James Laver's Time Spirit Theory, see also 1, 10, 30, 50, 70, 100 and 150 years*

The bio system for sewage to give trouble-free service for a residential dwelling—*this alternative to a septic tank works on the evapotranspiration principle*

The California incense cedar and the eastern hemlock trees to begin to flower

20y, 5mo
Life span of the tahr goat

20.9y
The average American female to marry for the first time

TWENTY YEARS TO TWENTY-FIVE YEARS

21y
To become a lieutenant colonel in the U.S. Army

The term of state supreme court judges in Pennsylvania

To reach the age at which delinquency codes are void and those charged with breaking the law are tried as adults in California and Wyoming

To reach the age when consent to proceed with surgery can be given without parental permission

Penicillin to come into widespread use after its discovery in 1928

To reach the legal age for entering into a contract in these states: Alabama (males), Arkansas (males), Colorado, Connecticut, Florida, Georgia, Idaho (male unless married, then the age is 18), Indiana, Louisiana, Maryland, Massachusetts, Minnesota, Mississippi, Missouri, Montana (males), Nevada (males), New Hampshire, New Mexico, New York, North Carolina, North Dakota (males), Ohio, Oregon, Rhode Island, South Carolina, Tennessee, Texas, Utah (males), Washington, Wyoming, and the District of Columbia

To reach the minimum age for holding office in the lower houses of most state legislatures

These varieties of trees to begin flowering: silver maple, sweetgum, white oak, and southern red oak

To reach the age of legal majority in these states: Alabama, Arkansas (males), Colorado, Connecticut, Florida, Georgia, Idaho (males), Indiana, Louisiana, Maryland, Massachusetts, Minnesota, Mississippi, Missouri, Montana (males), Nevada (males), New Hampshire, New Mexico, New York, North Carolina, North Dakota (males), Ohio, Oregon, Rhode Island, South Carolina, Tennessee, Texas, Utah (males), Wyoming, and the District of Columbia

21y, 5mo
Life span of the white-bearded gnu

21y, 9mo
Period for measuring the rule against perpetuities in most states

21y, 10mo
Life span of a coyote

22y
God to dictate the Koran to Mohammed

The stock market to recover from the Great Depression—*Although the crash came in 1929, the Dow Industrial average hit its low point in 1932 (41.22). It did not match the precrash 1929 high of 381.17 until 1954.*

To forget the names and faces of most high-school classmates after graduating—*according to a study at Ohio Wesleyan University*

Increased years of leisure that Americans enjoy—*compared with a generation ago*

To pay off the first half of a 30-year self-liquidating mortgage at 8% interest

Life span of the kinkajou

22.47y
Lifetime sleep requirement of the average American male

23y
To pay off the first one half of a 30-year self-liquidating mortgage at 10% interest

Life span of the panther

23.1y
Lifetime sleep requirement of the gorilla—*a gorilla sleeps 70% of the time*

Average American male to marry for the first time

To reach the age when polycystic kidney is usually detected

To get a cancerous tumor after extensive exposure to asbestos

23.3y
Average time of government service by members of Congress

23y, 8mo
Life span of the American badger

24y
To reach age of peak fertility in men and women

Religiously inhibited males to experience their first ejaculation—*Alfred C. Kinsey, University of Indiana study*

24.1y
Average term of FHA mortgages in 1950

24y, 3mo
Life span of the peccary

TWENTY-FIVE YEARS

Seventy-five percent of a group of men surviving heart attacks to complete their life span—*study of Metropolitan Life Insurance Company policy holders*

To become a full colonel in the U.S. Army

To become old enough to be elected as a U.S. representative

A car to be considered an antique—*and therefore eligible for special antique license tags*

The average adult to get out of purgatory—*estimate of Manly P. Hall in his published treatise* Death and After *(Hall Pub. 1929)*

Life expectancy of Gabonese men

Time between the discovery of liver damage as a result of working with vinyl chloride and the announcement of the first death due to cancer of the liver by B. F. Goodrich—*The discovery of liver damage due to working with vinyl chloride was made by a Russian scientist.*

To be considered a candidate for the Hall of Fame for Great Americans after death

Interval between celebrations of the Holy Year in the Roman Catholic church

To reach the silver wedding anniversary

Life span of the dwarf apple tree

Service requirement for retirement from the federal government—*or 20 years service at age fifty*

Interval between changes in design of U.S. coins

Growth of human bones to be complete—*longitudinal*

Maximum life span of the vicuna

The sugar maple to begin flowering

TWENTY-FIVE TO THIRTY YEARS

25y, 25d
Lifetime sleep requirement of the average American female

25y, 41d
Government employees to process the 43 million "data items" from the office of education for the state of Connecticut each year—*testimony of Connecticut school chief Mark Shedd before the Federal Paperwork Commission in January 1976*

25y, 2mo
Life span of the spotted hyena

25.9y
Career service of the average employee of the federal government

26y
Minimum time to become a surgeon—*premed program at university (4 years), med school (4 years), internship (1 to 2 years), residency (5 to 6 years)*

Schooling and experience to achieve professional status as an insurance actuary—*a college degree plus 5 to 10 years' experience*

American men to reach their full height in the 19th century

Life span of the Northern Pacific fur seal

26y, 3mo
Life span of the tiger

Life span of the cape buffalo

26y, 6mo
Life span of the red deer

27y
Life span of swine

27.9y
The comet Crommelin to orbit the Sun

28y
Life expectancy of women from Guinea

Life span of the cat

Life span of the California sea lion

28y, 4mo
Life span of the dromedary

29y, 5mo
Life span of a camel

29y, 5mo, 14d
Saturn to revolve around the Sun

29.9y
Average term of FHA mortgages in 1971

THIRTY YEARS

Range of childbearing years among American women
—*ages 15 through 44*

Life span of a termite

The accuracy of an ammonia resonator to vary one second

Dupont to develop Corfam

To reach the age when it is legal to run for office as a U.S. senator and for state senator in most states

The U.S. population to consume more minerals and mineral fuels than all the world previously—*1940–1970*

Life span of cows

The malarial parasitic cycle in the human body

The water tupelo tree to begin flowering

Fashionable clothes to be seen as "amusing"—*according to James Laver's Time Spirit Theory, see also 1, 10, 20, 50, 70, 100 and 150 years*

To reach the minimum age when the symptoms of Alzheimer's presenile psychosis appear (see also 40y)

THIRTY TO THIRTY-FIVE YEARS

30.5y
Career service of the average air-traffic controller

31y
Mesolithic Man to live his expected life span

33y
Maximum term of a rural housing loan from the Farmers Home Administration—*U.S. Department of Agriculture*

Life expectancy at birth for men in the Central African Republic

Half-life of the radio-isotope caesium 137

Expected life span of the Neanderthal Man

33y, 7mo
Maximum life span of the giraffe

34y
Maximum life span of the Mongolian wild horse

Maximum life span of the harbor seal

THIRTY-FIVE YEARS

To attain the age when a person is eligible to run for the office of the U.S. presidency

To develop the first radio

The period of time during which Ida May Fuller, the first person to receive U.S. Social Security, continued to get a monthly check—*Miss Fuller, who died on January 30, 1975 at the age of 100, began receiving her check on January 31, 1940. Miss Fuller contributed $22 to the fund before her retirement and received over $20,000 in benefits over the years*

To reach the coral wedding anniversary

The duration of life after the onset of multiple sclerosis symptoms—*U.S. Army study of WWII veterans*

For those who have inherited Huntington's disease to exhibit onset of symptoms

The gray seal to live its maximum life span

World population to double from 4 billion in 1975 to 8 billion

THIRTY-FIVE TO FORTY YEARS

The life expectancy for women in the Central African Republic

Copper age man (of west Turkey) to live out his expected life span

Life expectancy of men in ancient Greece

36.1y
Life expectancy of Liberian men

37y
To reach outer limit of bone growth in humans

The average span of years during which men can enjoy an active sex life

38.6y
Life expectancy of Liberian women

FORTY YEARS

Life span of the Indian rhinoceros

To reach the age when varicose veins become a likelihood

To reach the ruby wedding anniversary

The Israelites to wander in the wilderness—*Deut: 29*

The maximum term of a farm ownership loan from the Farmers Home Administration—*U.S. Department of Agriculture*

People to attain the age when glaucoma becomes a major health threat

Female buffalo to reach maximum reproductive age

To ring every possible combination of twelve cathedral change-bells—*full peal*

Bronze age man to live out his expected life span

The germination life of wild mustard, pigweed, ragweed, peppergrass and plantain seed—*they have germinated after being buried in 18 inches of soil accord-*

ing to J. D. Burnes, specialist at the University of Tennessee

To reach the maximum age for symptoms of Alzheimer's presenile psychosis to appear

FORTY TO FORTY-FIVE YEARS

40.6y
Life expectancy of Indian women

40.8y
Life expectancy for Burmese men

41y
Life expectancy for natives of Cameroon

41.9y
Life expectancy of Indian men

43y
The working life expectancy of Americans

43.3y
Life expectancy for Cambodian women

43.8y
Life expectancy for Burmese women

44y
The average life span of the American Indian

The Metropolitan Museum of Art to discover that a bronze horse exhibited as an example of ancient Greek sculpture was a fake, only 50 years old—*the introduction of gamma-ray shadow-graphs in the 1960s has been an invaluable aid in uncovering many art world fakes*

FORTY-FIVE YEARS

The life expectancy of Gabonese women

Eighteenth-century American men to live out their expected life span

The expected life span of a child at age six with cerebral palsy

The time between the discovery of the principle of the electric motor and the first practical use by Faraday

To reach the sapphire wedding anniversary

FORTY-FIVE TO FIFTY YEARS

47y
Life span of the horse

Stocks to change in value between the years 1928 and 1974:

Company	Value 12/31/74	Percent Change
Eastman Kodak	303,058	plus 2931
Proctor & Gamble	187,124	plus 1771
Sears Roebuck	151,216	plus 1412
United Aircraft	9,200	minus 8
Anaconda	4,928	minus 51
Woolworth	3,863	minus 61

47.5y
Life expectancy of Indonesian men and women

48y
The FBI to admit women to its ranks

48.3y
Life expectancy of Guatemalan men

49y
Men in Medieval England to live out their expected life span

49.7y
Life expectancy for Bolivian men and women and Guatemalan women

FIFTY YEARS

Life span of vultures

Life span of *clostridium botulinum* spores

The terms of loans to organizations from the Farmers Home Administration to protect watersheds—*U.S. Department of Agriculture*

To replace a forest destroyed by fire

The period that copyrights extend beyond the life of the author under the laws of Canada, Great Britain and U.S.

After Gutenberg invented the printing press for eight million books to be printed in fourteen countries

One fifth of the American population to reach the age when they lose all their teeth

An atom of neutral hydrogen to emit one 21-centimeter radio wave

To reach the golden wedding anniversary

Maximum time for giant sequoia tree to begin flowering (see also 15y)

Minimum time for untreated newspaper to decompose (see also 100y)—*depending upon climatic conditions*

Life expectancy of Chinese men and women

The world to run out of copper—*based on present estimated recoverable reserves at present mining and consumption rates*

Fashionable clothes to be seen as "quaint"—*according to James Laver's Time Spirit Theory, see also 1, 10, 20, 30, 70, 100 and 150 years*

FIFTY TO FIFTY-FIVE YEARS

51y
The President of the U.S. to get a private telephone in his office after phone service was installed in the White House—*until 1929 he used the booth in the hall*

Life span of the hippopotamus

Expected life span of European men in the seventeenth century

51.6y
Life expectancy for Egyptian men

52y
The life span of both Hermann the Great and Harry Houdini—*pointed to as significant by numerologists who claim both magicians' names carry the vibratory influence of "5" associated with travel and adventure*

and the total vibration of "1" indicating success. Both died suddenly in the prime of their careers.

A calendar cycle by Mayan reckoning

Life expectancy of Jordanian women

53.8y
Life expectancy for Egyptian women

54y, 31d
Two eclipses to be seen on the same spot of the Earth

54.4y
Life expectancy of Chilean men

FIFTY-FIVE TO SIXTY YEARS

55y
To reach the age of compulsory retirement in Japan

To reach the emerald wedding anniversary

55.5y
Life expectancy of Peruvian women

56y
To produce a white marigold—*Burpee*

Period of validity of a U.S. registered trademark

56.6y
Life expectancy of men from El Salvador

57.2y
Life expectancy of men from the Dominican Republic

58y
The population of North America to double—*estimate beginning 1971*

58.6y
Life expectancy of women from the Dominican Republic

58.7y
Life expectancy of Thai women

59y
Life expectancy of men from Guyana

59.5y
To reach the age when the federal government allows the self-employed to use Keogh plan money for retirement income (see also 70.5y)

59.7y
Life expectancy of South Korean men

59.9y
Life expectancy of Chilean women

SIXTY YEARS

Life span of the African elephant

Seeds to lose their germination potency in a sealed jar

A calendar cycle by Oriental reckoning—*both Hindu and Chinese*

Maximum period of sexual potency among males

Expected life span of American men at the turn of the twentieth century

The first life span of ma-dake bamboo—*after maturing, flowering, and dying, it lies dormant for 15 years before new shoots grow from underground stems*

SIXTY TO SIXTY-FIVE YEARS

60.4y
Life expectancy of women from El Salvador

60.9y
Life expectancy of Panamanian women

Life expectancy of Mexican men

61.4y
Life expectancy of women in Ceylon

61.9y
Life expectancy of men in Ceylon and Costa Rica

62.2y
Life expectancy of men in Trinidad and Tobago

62.7y
Life expectancy of Jamaican men

63y
Life expectancy of Navajo Indians

Life expectancy of women from Guyana

63.7y
Life expectancy of Mexican women

64.1y
Life expectancy for Argentine men and South Korean
women

64.3y
Life expectancy of Yugoslavian men

64.8y
Life expectancy of Costa Rican women

64.9y
Life expectancy for Albanian men

SIXTY-FIVE TO SEVENTY YEARS

65y
Life expectancy of men in the Soviet Union

Average life span of the black locust tree

Life expectancy of an Australian Aborigine

Age at which osteoporosis begins to be a major health
threat

65.3y
Life expectancy of Portuguese men

65.4y
Life expectancy of Finnish men

65.5y
Life expectancy of Rumanian men

65.8y
Life expectancy of Nationalist Chinese men

66.3y
Life expectancy of women born on Trinidad and
Tobago and men in Austria

66.6y
Life expectancy of Hungarian men and Jamaican
women

66.9y
Life expectancy of Polish and Scottish men

67y
Life expectancy for Albanian women

67.3y
Life expectancy of Czechoslovakian and Spanish men

67.4y
Life span of the average American male

67.5y
Life expectancy for Greek men

67.6y
Life expectancy of West German and French men

67.7y
Life expectancy for Belgian men

67.9y
Life expectancy of men born in Northern Ireland, Italy, and Australia

68.1y
Life expectancy of Irish men

68.4y
Life expectancy of male New Zealanders and Maltans

68.6y
Life expectancy of men in England and Wales

68.7y
Life expectancy of Swiss men

68.8y
Life expectancy of Bulgarian and Canadian men

68.9y
Life expectancy for Yugoslavian women

69.1y
Life expectancy of Japanese men

69.2y
Life expectancy of East German men

69.6y
Life expectancy of Israeli men

The comet Olbers to orbit the Sun

69.8y
Life expectancy of Rumanian women

SEVENTY TO SEVENTY-FIVE YEARS

70y
Life span of the Indian elephant

Twenty-seven percent of the male population to reach the age when they become impotent

Fashionable clothes to be seen as "charming"—*according to James Laver's Time Spirit Theory (see also 1, 10, 20, 30, 50, 100 and 150 years)*

For per capita sugar consumption to increase in the United States 120%—*1902–1972*

Maximum germination life of curley dock, mullein, and primrose seed—*These seeds have surfaced and germinated after this period in the ground according to J. D. Burnes, specialist at the University of Tennessee.*

70.1y
Life expectancy of Danish men

70.2y
Life expectancy of Argentine women

70.4y
Life expectancy of Nationalist Chinese women

70y, 6mo
Self-employed individuals to reach the age when they must begin receiving benefits from their tax-deferred, Keogh retirement plan—*they need not retire however*

70.7y
Life expectancy of Dutch men and Greek women

70.8y
Life expectancy of men in Iceland

70.9y
The comet Pons-Brooks to orbit the Sun

71y
Life expectancy of Norwegian men and Portuguese women

71.9y
Life expectancy of Spanish, Hungarian, and Irish women and of Swedish men

72y
To reach the age when a person can have an unlimited income without forfeiting any Social Security benefits

Estimated time to build the National Cathedral in Washington, D.C.—*begun in 1908, it is scheduled for completion in 1980*

72.6y
Life expectancy of Finnish and Maltan women

72.7y
Life expectancy of Bulgarian women

72.8y
Life expectancy of Polish women

Life expectancy of Israeli women

73.1y
Life expectancy of Scottish women

73.4y
Life expectancy of Italian women

73.5y
Life expectancy of women born in Northern Ireland,
Belgium, and Austria

73.6y
Life expectancy of West German and Czechoslovakian
women

73.8y
Life expectancy of women in New Zealand

74y
Life expectancy of women in the Soviet Union

74.1y
Life expectancy of Swiss women

74.2y
Life expectancy of Australian women

74.3y
Life expectancy of Japanese women

74.4y
Life expectancy of East German women

74.9y
Life expectancy of women born in England and Wales

SEVENTY-FIVE TO EIGHTY YEARS

75y
Life span of carp (see also 15y)

To reach the diamond wedding anniversary

75.2y
Life expectancy of Canadian women

Life expectancy of American women

75.3y
Life expectancy of French women

75.6y
Life expectancy of Danish women

76y
Life expectancy of Norwegian women

76.2y
Life expectancy of women in Iceland

76y, 2mo, 13d
Halley's comet to orbit the Sun

76.5y
Life expectancy of Swedish and Dutch women

77y
Life expectancy of a man adhering to the strict dietary laws of the Seventh-Day Adventists—*study by Loma Linda University*

EIGHTY YEARS

The world's population to double from one billion in 1850 to two billion (see also 35–37y; 45y; 200y; 1,650y)

To become an octogenarian

Life expectancy of a woman adhering to the strict dietary laws of the Seventh-Day Adventists

EIGHTY-FIVE TO NINETY-FIVE YEARS

85y
Manhattan to be rediscovered by Hendrick Hudson after the first known white men sailed into the bay in 1524—*they were Italian explorers Giovanni and Hiaro Verrazano*

88y
The population of Europe to double—*estimate beginning 1971*

91y
Between the discovery of vinyl chloride gas by Regnault in 1835 and discovery of the process turning the gas into polyvinyl chloride by Dr. Waldo Semon of B.F. Goodrich

93y
A dozen escaped gypsy moths to multiply to a number which defoliated 2 million acres of northeastern forest —*in the United States in 1972*

NINETY-FIVE YEARS

Easter to recur on the same date

ONE HUNDRED YEARS

A century

Life span of the southern poplar

Tidal friction to slow the Earth's rotation by 14 seconds

Furniture to become antique—*according to U.S. Customs guidelines which allow its duty-free entry into the country*

The Earth's ozone layer to be replenished after molecular destruction by fluorocarbons released at the Earth's surface from aerosol cans (see also 10y)

The population of the United States, Western Europe, and the Soviet Union to reach one and a half billion and the rest of the world to reach four billion

The leaning tower of Pisa to increase its inclination by 1 foot

Members of the Tredwell family to inhabit 29 E. 4th Street, New York City—*When the last residing family member died in 1933, the house, known as Old Merchant's House, was turned into a museum, preserved just as it was, chronicling the life of America's middle-class family for a century. (The only family to ever occupy it.)*

Maximum time in best climatic conditions for untreated newspaper to decompose (see also 50y)

Portions of the Atlantic seaboard and coast of Holland to sink about a foot

For the Earth's mean solar day to be shortened by .001 of a second—*due to the slowing of the Earth's rotation*

A clock which kept perfect time and which was perfectly synchronized with solar time in the year 1750 to become 2 seconds slow—*due to the fact that the Earth's rotation is slowing down*

The light emitting diodes on a pulsar watch to lose their brightness with continuous use

Fashionable clothes to be seen as "romantic"—*according to James Laver's Time Spirit Theory (see also 1, 10, 20, 30, 50, 70, and 150 years)*

Average life span of the savin juniper

110y
The human body to wear out

Average life span of the red maple

112y
Average life span of the northern catalpa and dogwood trees

120y
Average life span of the American linden tree

Moses to live his life

125y
The world to run out of nickel—*based on present estimated recoverable reserves at present mining and consumption rates*

Average life span of the American holly and the common hockberry trees

126y
To build St. Peter's Church in Rome

128y
The Julian Calendar to gain a full day on the Sun

Average life span of the black cherry tree

U.S. Naval Academy to accept women—*1975*

140y
The populations of Sweden, Denmark, and the United Kingdom to double—*estimate beginning 1971*

150y
To reach a sesquicentennial

Fashionable clothes to be seen as "beautiful"—*according to James Laver's Time Spirit Theory (see also 1, 10, 20, 30, 50, 70 and 100 years)*

164y, 9mo, 19d
Neptune to make one revolution around the Sun

175y
The world to run out of manganese, cobalt, and aluminum—*based on present estimated recoverable reserves at the present mining and consumption rates*

Average life span of the loblolly pine

Abraham to live his life

TWO HUNDRED YEARS

A bicentennial

Average life span of the slash pine, pacific yew, eastern black walnut and Arizona cypress

224y
Jail sentence received by Spanish political figure Vila Reyes in May 1975 for fraud—*His lawyer said he would probably serve 15 years*

225y
Average life span of the shag back hickory, the grand fir tree, the longleaf pine and American elm

232y
Sotheby's to break its all-male auctioneer rule—*Libby Howie, 24 years old, broke it in 1976*

235y
Average life span of the shortleaf pine

248.4y
Pluto to revolve around the Sun

250y
Life span of modern movie film with careful storage —*Current films are treated with acetic acid compared with the early nitric acid treated stock which has disintegrated in less than 40 years.*

262y
Average life span of the white fir tree

265y
To bore a hole 18 inches in diameter through the center of the Earth—*as soon as an earth-boring machine big enough is built*

275y
Average life span of the shore pine, the sugar maple, the cascades fir, American beech, white ash and the red spruce

THREE HUNDRED YEARS

Average life span of the western white pine

Life span of a mermaid—*from* The Little Mermaid *by Hans Christian Andersen*

325y
Average life span of the western larch

350y
Average life span of the California incense cedar, the eastern hemlock, and the western hemlock, and the red fir tree

FOUR HUNDRED YEARS

400y
Average life span of the noble fir, Lawson's fake cypress, and the sugar pine

From the time Leonardo da Vinci made sketches of gliders to the first real airplane

The cycle during which February contains five Sundays: 3 times every 28 years; then once after 40 years; then 2 times after 28 years; and once after 40 years; followed by 3 times at 28 years—*for a total of 400 years, when the cycle begins again*

A quadricentennial

425y
The world to run out of iron—*based on present estimated recoverable reserves at present mining and consumption rates*

450y
Average life span of the ponderosa pine and the white oak

FIVE HUNDRED YEARS

525y
The world to run out of chromium—*based on present estimated recoverable reserves at present mining and consumption rates*

550y
The United States to run out of coal reserves estimated at 3.2 trillion tons

The Moon's shadow to strike any spot on the Earth twice

The life span of the Sitka spruce

584y
A one billion years' supply of a natural resource to be exhausted if the rate of present consumption is increased by 3% each year—*Vincent McKelvey, director of the U.S. Geological Survey*

SIX HUNDRED YEARS

600y
The radioactive nuclear wastes, strontium-90 and cesium-137, to become harmless

Average life span of the bald cypress

650y
Average life span of the nootka false cypress

Oxford to admit women to its degree program

675y
Average life span of the giant arborvitae

SEVEN HUNDRED YEARS

750y
Average life span of the Douglas fir tree

NINE HUNDRED YEARS

The chalk cliffs of Dover to erode two miles—*since the time of the Norman Conquest in 1066*

930y
Adam, father of us all, to live his life

950y
Noah to live his life

969y
Methuselah to live his life

ONE THOUSAND YEARS

A millenium

A migrating people to lose 20% of their mother language—*Glottochronology, the linguistic science which chronicles this phenomenon, was proposed by M. Swadesh in 1950 as a method of deducing, on the basis of statistical comparisons, family relationships of languages as well as the probable date when branches of a given language group or family separated from the common parent language.*

The United States to consume the 400 billion tons of recoverable coal contained within its boundaries

1,100y
Average life span of the redwood tree

1,461y
Sirius to rise again with the Sun according to the Egyptian calendar—*Sothic cycle*

1,600y
A real steam engine to be developed after the ancient Egyptians introduced toys with movable parts run by steam

1,622y
Half-life of radium

1,800y
To complete the Great Wall of China—*Begun in the third century* B.C., *it received its last addition in the sixteenth century* A.D.

1,850y
The world's population to grow from 200 million in the year A.D. 1 to 1 billion

TWO THOUSAND YEARS

Interval between glass blowing and the first machine-made glass bottles which made mass production possible—*Glass blowing was discovered in the first century but it wasn't until 1903 that mass production of glass bottles took place.*

Indian lotus seeds to lose their ability to germinate

A pulsar watch to wear out if the time is checked 25 times a day

2,500y
Average life span of the giant sequoia

2,700y
Period mummies remain intact—*one this old has been disected recently*

THREE THOUSAND YEARS

3,000y
A banto peach to ripen—*This legendary Japanese fruit bestows health and everlasting life on those lucky enough to taste it.*

The Gregorian Calendar's variance with the Sun to total 1 day

3,400y
The Gregorian Calendar to gain 1 day on the Sun

3,500y
The telescope and microscope to be developed after the invention of the lens

FIVE THOUSAND YEARS

The Bronze Age

5,400y
Our current North Star, Polaris, to be replaced by Alpha Cephei, as the sideway motion of the Earth's axis begins shifting away from Polaris after the year 2100

5,600y
Half-life of carbon 14—*used for dating organic matter*

SEVEN THOUSAND YEARS

7,980y
The Julian period—*first year was 4713 B.C.*

TEN THOUSAND YEARS

Between ice ages

Technological developments of civilization to rend a planet uninhabitable—*theory of L. E. Orgel*

12,000y
The North Pole to shift alignment from Polaris to the star Vega as the Earth's axis shifts—*Polaris is our current North Star*

18,000y
The Chinese god P'an Ku to create the universe (see also 6d)

TWENTY THOUSAND YEARS

24,000y
The half-life of plutonium

25,800y
The axis of the North Pole to make a complete circle (precession) and for the solstices and equinoxes to make a complete circuit of the skies (also called the Great Year)

28,000y
Homo sapiens to become agrarian

THIRTY THOUSAND YEARS

30,000y
The accuracy of a cesium resonator to vary one second

FORTY THOUSAND YEARS

46,720y
The seasons to pass through the entire year—*according to the Julian Calendar*

FIFTY THOUSAND TO ONE HUNDRED THOUSAND YEARS

75,000y
Comet Kohoutek to complete its orbit so it will again be visible to observers on Earth

85,000y
Neanderthal man to be replaced by the Cromagnon man

90,000y
Java man to be replaced by the Neanderthal man

100,000y
Light to travel across our galaxy

ONE HUNDRED THOUSAND TO ONE MILLION YEARS

370,000y
Range of accurate dating possible under the ancient Mayan system of time reckoning

500,000y
The radioactive nuclear waste plutonium 239 to become harmless

ONE MILLION TO FIFTY MILLION YEARS

1,000,000y
Quaternary period of the Cenozoic Era

2,000,000y
Light to reach the Earth from Andromeda Galaxy

The Colorado River to carve the Grand Canyon

Light to travel to us from the closest neighboring galaxy

2,500,000y
Before the next occurrence of a February without a full Moon—*the last occurred in 1866*

2,990,000y
The world's largest ground sloth to reach extinction during the Pleistocene age

3,000,000y
The accuracy of the hydrogen maser to vary one second

A 30-thousand-foot-thick layer of lava to cool in the interior from 1,100° C. to 750° C.

6,000,000y
The Eocene and Paleocene epochs of the Tertiary period of the Cenozoic Era

12,000,000y
The Miocene epoch during the Tertiary period of the Cenozoic Era

13,000,000y
The Pliocene epoch during the Tertiary period of the Cenozoic Era

20,000,000y
The Silurian period of the Paleozoic Era

30,000,000
The Pennsylvania period of the Paleozoic Era

35,000,000y
The Mississippi period of the Paleozoic Era

42,000,000y
The Tertiary period during the Cenozoic Era

43,000,000y
The Cenozoic Era

46,000,000y
The Jurassic period during the Mesozoic Era

49,000,000y
The Triassic period during the Mesozoic Era

50,000,000y
The Permian period of the Paleozoic period

The Sun to experience gravitational collapse during its formation

FIFTY MILLION TO ONE HUNDRED MILLION YEARS

60,000,000y
The Devonian period of the Paleozoic Era

70,000,000y
South America and Africa to drift 2,000 miles apart—
*135 million years ago to 65 million years ago, accord-
ing to the theory of continental drift first postulated
60 years ago by German Alfred Wegener*

72,000,000y
The Cretaceous period during the Mesozoic Era

75,000,000y
The Ordovician period of the Paleozoic Era

76,000,000y
The Earth's magnetic field to reverse at least 171 times

100,000,000y
The Cambrian period of the Paleozoic Era

A neutron star to cool

ONE HUNDRED MILLION TO ONE BILLION YEARS

230,000,000y
The Sun to make one rotation within its galaxy

250,000,000y
Petroleum to form (see also 1y)

300,000,000y
Before we can expect to have a bad nuclear accident
—*if history follows the odds. The Atomic Energy*

Commission's Dixy Lee Ray feels the chance of nuclear power plants being responsible for mass destruction is about as remote as a giant meteor hitting the Earth, once in 300 million years.

The Appalachian Mountains to form

370,000,000y
The Paleozoic Era

ONE BILLION TO TWO BILLION YEARS

1,000,000,000y
Uranium to radiate as much energy as 5,000 pounds of burning gasoline

Light to reach the Earth from the most distant bodies ever photographed

For life to evolve on Earth after material to sustain life became available

The Sun to emit as much energy as a supernova releases in 24 hours—*stellar explosion*

At its present rate of change, the Earth to rotate only once a year—*the same side of the Earth will be turned continually toward the Sun*

Coal to form (see also 1y)

1,300,000,000y
Half-life of potassium 40

2,000,000,000y
Light from the Hydra cluster to reach the Earth

The gaseous universe to cool enough following crea-
tion for stars to be formed—*theory of California
Institute of Technology astronomer Allan Sandage*

TWO BILLION TO FOUR BILLION
FIVE HUNDRED MILLION YEARS

Iron to form (see also 1y)

2,500,000,000y
The Precambrian Era

4,000,000,000y
The force of gravity of the Moon upon the Earth to
be halved—*so that people and other objects on the
Earth weigh half as much as they did when the Earth
was young*

Lead to form (see also 1y)

4,500,000,000y
The planet Earth to arrive at its present state

The half-life of uranium

FIVE TO TEN BILLION YEARS

The Sun to become a "red giant"—*when its supply
of hydrogen begins to run low*

EIGHT BILLION YEARS

Light from a quasi-stellar radio source to reach the
Earth—*oldest known light source*

TEN BILLION YEARS

Before the Earth is no longer able to support life due to the impending death of the Sun

The average life span of a planet

BIBLIOGRAPHY

Aaronson, Charles S., ed., *International Motion Picture Almanac*, New York, Quigley, 1967.

Alexander, Conel Hugh O'Donel, *A Book of Chess*, New York, Harper & Row, 1973.

Altman, Philip L., and Dittmer, Dorothy S., *Biological Data Book* 3 Vol., Bethesda, Md. Federation of American Societies for Experimental Biology, 1972.

Arlin, Marian Thompson, *The Science of Nutrition*, New York, Macmillan, 1972.

Asimov, Isaac, *The Clock We Live On*, New York, Abelard Schuman, 1959.

Asimov, Isaac, *Left Hand of the Electron*, Garden City, N.Y., Doubleday, 1972.

Asimov, Isaac, *The Solar System and Back*, Garden City, N.Y., Doubleday, 1970.

Bailey, Gerald B. and Presgrave, Ralph, *Basic Motion Time Studies*, New York, McGraw-Hill, 1958.

Barnes, Ralph Mosser, *Motion & Time Study*, New York, John Wiley & Sons, 1963.

Bendann, Effie, *Death Customs*, New York, Knopf, 1930.

Bent, Arthur Cleveland, *Life Histories of North American Birds*, 3 Vol., Washington, D.C., Smithsonian Institute Press, 1968.

Better Homes and Gardens New Cookbook, New York, Meredith Press, 1968.

Better Homes and Gardens New Garden Book, New York, Meredith Press, 1968.

Blair, Thomas Arthur, *Weather Elements*, Englewood Cliffs, N.J., Prentice-Hall, 1965.

Booth, Verne Hobson, *Physical Science*, New York, Macmillan, 1967.

Brooks, William Osbert et. al., *Modern Physical Science*, New York, Holt, Rinehart & Winston, 1970.

Bullough, W. S., *Vertebrate Sexual Cycles*, London, Methuen & Co., 1951.

Burnes, Eugene, *Sex Life of Wild Animals*, Toronto, Clarke Irwin, 1953.

Buros, Oscar Krisen, ed., *Mental Measurements Yearbook*, Highland Park, N.J., Gryphon Press, 1972.

Burton, Maurice, ed., *The Encyclopaedia of Animals*, London, Octopus Books; Sydney, Angus & Robertson, 1972.

Calder, Peter Ritchie, *Evolution of the Machine*, New York, American Heritage Press, 1968.

Calvert Party Encyclopedia, Calvert Distillers Co., 1964.

Candland, Douglas K., *Psychology, The Experimental Approach*, New York, McGraw-Hill, 1968.

Cayne, Bernard S., ed., *Encyclopedia Americana*, New York, Americana Corporation, 1975: "Artificial Respiration," Gordon, Archer S., Vol. 2, pg. 416; "Cattle" Vol. 6, pg. 77; "Childbirth" Greenhill, J. P., Vol. 6, pg. 470; "Insect" Oldroyd, Harold, Vol. 15, pg. 202; "Magellan" Vol. 18, pg. 115.

A Century of Wonder, Popular Science Monthly, New York, Doubleday, 1972.

Clark, Merrian E., ed., *Ford's Freighter Travel Guide*, Woodland Hills, Ca., Clark, 1974.

Clark, Randolph L., *The Book of Health*, New York, Van Nostrand, Reinhold, 1973.

Clark, Sydney Alymer, *All the Best in Europe*, New York, Dodd Mead, 1967.

Cowan, Harrison J., *Time and Its Measure*, Cleveland, Ohio, World Publishing, 1958.

Crowther, James Gerald, *Discoveries & Inventions of the 20th Century*, London, Routledge & Kegan, 1966.

Culture In America, Washington, D.C., *Congressional Quarterly*, 1969.

DuBoise, J. Harry, *Plastics*, New York, Holt, Rinehart & Winston, 1969.

Durin, John Valentine George Andrew and Passmore, Reginald, *Energy, Work and Leisure*, London, Heinemann Education Books, Ltd., 1967.

Editorial Research Reports on Cultural Life and Leisure in America, Washington, D.C., *Congressional Quarterly*, 1969.

Ehrlich, Paul, *The Population Bomb*, New York, Ballentine, 1971.

Encyclopedia of Textiles, American Fabric Magazine, Englewood Cliffs, N.J., Prentice-Hall, 1972.

The Family Circle Cookbook, food editors of *Family Circle* and Jean Anderson, New York, *Family Circle*, 1974.

Ferretti, Frederick, *Great American Marble Book*, New York, Workman Pub., 1973.

Ford Motor Company Flat Rate Manual, 1974.

Frontiers of Psychological Research (Readings from the Scientific American), San Francisco, W. H. Freeman and Co., 1966.

Gabrielson, Ira Noel, *New Fisherman's Encyclopedia*, Harrisburg, Pa., Stackpole Books, 1963.

Gagnier, Ed, *Inside Gymnastics*, Chicago, Regnery, 1974.

Garrett, Henry Edward, *Great Experiments in Psychology*, New York, Appleton-Century-Crofts, 1951.

Gauger, William, "Household Work, Can We Add It to the GNP?", *Journal of Home Economics*, October 1973, pg. 12-15.

Gelatt, Roland, *The Fabulous Phonograph*, New York, Appleton-Century-Crofts, 1965.

Gibson, Walter B., *The Science of Numerology, What the Numbers Mean to You*, New York, G. Scully, 1927.

Godfrey, Robert Sturgis, ed., *Means Building Construction Cost Data*, Duxbury, Ma., Robert Snow Means Co., 1974.

Goodman, Louis S. and Gilman, Alfred, *The Pharmacological Basis of Therapeutics*, New York, Macmillan, 1970.

Gord, Michael, *Time and Moneysavers in the Kitchen*, New York, Dolphin Books, 1966.

Gove, Philip Babcock, ed., *Webster Third International Dictionary*, Springfield, Ma., Merriam, 1968.

Graham, Frank D., *Questions and Answers for Engineer's and Fireman's Examinations*, New York, Audel, 1974.

Gregorian Calendar

Grossman, H. J., *Grossman's Guide to Wines, Spirits and Beers*, New York, Charles Scribner's Sons, 1964.

Hall, Manley P., *Death and After*, Los Angeles, Hall, 1929.

Hall, Florence T. and Schroeder, Marguerite D., "Time Spent on Household Tasks," *Journal of Home Economics*, January 1970, pg. 23-29.

Hammer, Philip G. and Chapen, F. Stuart, *Human Time Allocation: A case study of Washington, D.C.*, Chapel Hill, N.C., Center for Urban and Regional Studies, University of North Carolina, 1972.

Hammesfahr, James E. and Stong, Clair L., *Creative Glass Blowing*, San Francisco, W. H. Freeman, 1968.

Hepburn, Andrew, *Rand McNally Guide: New York City*, Chicago, Rand, 1970.

Hertberg, Hendrik, *One Million*, New York, Simon and Schuster, 1970.

Hoblitzelle, Lucy F. and Winek, Charles L., *Pharmacology Applied to Patient Care*, Worcester, Ma., Davis, 1969.

Hoehn, Robert G., *Illustrated Guide to Individual and Team Baseball*, Nyack, N.Y., Parker, 1974.

Holvey, David N., ed., *Merck's Index*, West Point, Pa., Merck Sharp and Dohme Research Laboratories, 1972.

Holy Bible, Authorized King James Version, New York, Oxford University Press, n.d.

HUD Statistical Yearbook, Washington, D.C. GPO, 1971.

Hunt, Morton M., *Sexual Behavior in the 70's*, Chicago, Playboy Press, 1974.

Jacobs, Harold, *Mathematics, A Human Endeavor*, San Francisco, W. H. Freeman, 1970.

Jarman, Catherine, *Atlas of Animal Migration*, London, Heinemann, 1972.

Jennison, Keith Warren, *The Concise Encyclopedia of Sports*, New York, Franklin Watts, 1970.

Jessup, Libby, *How to Become a Citizen of the U.S.*, Dobbs Ferry, N.Y., Oceana Publications, 1959.

Kaye, Evelyn, *The Family Guide to Children's Television*, New York, Pantheon, 1974.

Kendall, Lace, *Tigers, Trainers and Dancing Whales*, Philadelphia, Macrae Smith, 1968.

Kennedy, Edward R., ed., *1974 World Almanac and Book of Facts*, New York, Newspaper Enterprise Association.

Kibbe, Constance, *Standard Textbook of Cosmetology*, New York, Milady, 1967.

Kingsbury, John M., *Poisonous Plants of the United States and Canada*, Englewood Cliffs, N.J., Prentice-Hall, 1964.

Kinsey, Alfred Charles, et al., *Sexual Behavior in the Human Male; Sexual Behavior in the Human Female*. University of Indiana, 1968.

Koester, Jane and Adkins, Rose, ed., *Writer's Market 1974*, Cincinnati, Ohio, Writer's Digest, 1973.

Koppet, Leonard, *New York Times Guide to Spectator Sports*, New York, Quadrangle, 1971.

Luce, Gay, *Biological Rhythms in Humans and Animal Physiology*, New York, Dover, 1971.

Lugo, James O. and Hershey, Gerald L., *Human Development*, New York, Macmillan, 1974.

Manaka, Yashio and Urquhart, I. A., *Layman's Guide to Acupuncture*, New York and Tokyo, Weatherhill, 1972.

Mann, Martin, ed., *Time-Life Science Library*, 26 Vol., New York, Time-Life Books, 1969.

Masters, William H. and Johnson, Virginia E., *Human Sexual Inadequacy*, Boston, Little, Brown, 1970.

McClurg/Shoemaker, Architects and Engineering, ed., *The Building Estimator's Reference Book*, Chicago, Frank R. Walker, 1970.

McCully, Helen, *Nobody Ever Tells You These Things*, New York, Holt Rinehart & Winston, 1967.

McGinnis, R. J., *The Good Old Days*, New York, Harper & Row, 1960.

The Medicine Show, ed. of Consumer Reports, Mt. Vernon, N.Y., Consumers Union, 1970.

Menke, Frank Grant, *Encyclopedia of Sports*, Cranbury, N.J., A. S. Barnes & Co., 1963.

Mitford, Jessica, *The American Way of Death*, New York, Fawcett World, 1973.

Moore, Francis Daniels, *Transplant, The Give and Take of Tissue Transplantation*, New York, Simon & Schuster, 1972.

Moore, John L., ed., *Continuing Energy Crisis in America*, Washington, D.C., Congressional Quarterly, 1973.

Nobile, Philip and Deedy, John, *The Complete Ecology Fact Book*, New York, Doubleday, 1972.

Occupational Outlook, Handbook of the U.S. Bureau of Labor Statistics, Washington, D.C. GPO, 1974.

Ohlin, Lloyd, *Prisoners in America*, Englewood Cliffs, N.J., Prentice-Hall, 1973.

Oliver, John W., *History of American Technology*, New York, Ronald Press, 1956.

Orgel, Leslie E., *The Origins of Life*, New York, Wiley & Sons, 1973.

Our American Government, House Document 93-153, Washington, D.C. GPO, 1973.

Passel, Peter and Ross, Leonard, *The Best*, New York, Farrar Strauss and Giroux, 1974.

Physician's Desk Reference to Pharmaceutical Specialties, Oradell, N.J., Medical Economics Co., Litton Div., 1974.

Pszczola, Lorraine, *Archery*, Philadelphia, W. B. Saunders Co., 1971.

Purdy, Susan Gold, *Jewish Holidays*, Philadelphia, Lippincott, 1969.

Reader's Digest Almanac, Pleasantville, N.Y., 1973.

Recommended Dietary Allowances, Publication 1694, Dietary Allowance Commission and Food and Nutrition Board, Washington, D.C., National Academy of Science, National Research Council, 1974.

Redding, William J., ed., *Lincoln Library of Essential Information*, Columbus, Ohio, Frontier Press, 1970.

Roast, Harold J., *Cast Bronze*, Metals Park, Novelity, Ohio, American Society for Metals, 1953.

Rosenberg, Jerome Laib, *Photosynthesis*, New York, Holt Rinehart & Winston, 1965.

Rothenberg, Robert Edward, *Understanding Surgery*, New York, Pocketbooks, 1973.

Rowe, Mary, "The Length of a Housewife's Day in 1917" (fr. Dec. 1917 *Journal of Home Economics*, pg. 16-19) October 1973 *Journal of Home Economics*, pg. 7-11.

Sandow, Stuart A. with Kelly, Edward F., *Components of a "Futures Perspective,"*: *An Exploratory Study*, Syracuse, N.Y., Syracuse University Research Corp., 1973.

Sandow, Stuart A., "You Can't Tell Which Way the Train Went Just by Looking at the Tracks" privately published by author, 1973.

Schwartz, Alvin, *America's Exciting Cities*, New York, Crowell, 1966.

Sharp, Samuel L., *The Soviet Union and Eastern Europe*, Washington, D.C., Stryker-Post, 1973.

Smart, Mollie S. and Smart, Russell C., *Children, Development and Relationships*, New York, Macmillan, 1972.

Statistical Abstract of the United States, U.S. Department of Commerce, Bureau of Census, Washington, D.C. GPO, 1973.

Steinlage, Gerald, F., *Wines Brewing Distillation*, St. Henry, Ohio, Steinlage Products, 1972.

Still, Henry, *Of Times, Tides and Inner Clocks*, Harrisburg, Pa., Stackpole Books, 1972.

Stokes, William, *Essentials of Earth History*, Englewood Cliffs, N.J., Prentice-Hall, 1966.

Swezey, Kenneth Malcolm, *Formulas, Methods, Tips and Data for Home and Workshop*, New York, Harper & Row, 1971.

Szalai, Alexander, ed., in collaboration with Converse, Philip E. et al., *The Use of Time, Daily Activities of Urban and Suburban Populations in 12 Countries*, The Hague, Mouton, Report of Study by the European Coordination Center for Research and Documentation in Social Sciences, 1973.

Tucker, Richard K. and Crabtree, D. Glen, *Handbook of Toxicity of Pesticides to Wildlife*, Department of Interior, U.S. Bureau of Sports, Fisheries and Wildlife, Washington, D.C. GPO, 1970.

Tufty, Barbara, *1001 Questions Answered About Storms*, New York, Dodd, Mead, 1970.

Tyarks, Frederic Ewald, *Europe on a Shoestring*, Greenlawn, N.Y., Harian, 1973.

Tyler, Hamilton, *Organic Gardening*, New York, Van Nostrand, Reinhold, 1970.

The World and Its People: Middle East, 2 Vol., New York, Greystone Press, 1967.

United Nations Statistical Yearbook, New York, UN, 1974.

U.S. Federal Codes, Indianapolis, Ind., Bobbs-Merrill Co., 1968.

Waldo, Myra, *Travel and Motoring Guide of Europe*, New York, Macmillan, 1973.

Walker, Kathryn E., "Household Work Time, Its Implications for Family Decisions," *Journal of Home Economics*, October, 1973.

Whitaker, Joseph, *Whitaker's Almanack*, London, William Clowes and Sons, Ltd., 1974.

White, William, *Guppy, Its Life Cycle*, New York, Sterling, 1974.

Yeates, M. H. and Garner, B. J., *The North American City*, New York, Harper & Row, 1971.

You and the Law, Pleasantville, N.Y., Reader's Digest, 1971.

Zimmerman, Joseph Francis, *State and Local Government*, New York, Barnes & Nobel, 1970.

PERIODICALS

The American City
Cathedral Age
Consumers Report
Education Daily
Family Circle
Flower and Garden
Johnny Horizon
Good Housekeeping
Meyer Seed Catalog
The Farmers' Almanac
Hobbies
Money
Motor Trend
Moneysworth
National Geographic

Newsweek
The New York Times
Parade
Popular Electronics
Popular Science
Psychology Today
The Rolling Stone
Sports Illustrated
Time
Trains
The Washington Post
Womensports
U.S. Naval Institute Proceedings
Vogue

PUBLICATIONS AND PUBLIC
INFORMATION OFFICES

American Automobile Association
American Bar Association
American Cancer Society
American Medical Association
Chesapeake and Potomac Telephone Company
Children's Defense Fund
Council of Family Health
Dairy Council of the Upper Chesapeake Bay
Environmental Protection Agency
Federal Bureau of Investigation
First National City Bank of New York
Howard County (Maryland) Office of Civil Defense
International Association of Fire Fighters
Law Enforcement Assistance Administration
Library of Congress Talking Books for the Blind
Maryland State Department of Motor Vehicles
Metropolitan Life Insurance Co.
Motor Vehicle Manufacturers Assn. of U.S.
National Archives
National Broadcasting Company
National Fire Protection Association
Population Reference Bureau
Public Broadcasting System
United Services Automobile Association
U.S. Department of Agriculture
U.S. Bureau of the Census
U.S. Department of Commerce
U.S. Bureau of Labor Statistics
U.S. Food and Drug Administration
U.S. Office of Management and the Budget
U.S. Postal Service

CONSULTANTS

Elaine Y. Bicksler, quilt maker
Mary Catherine Bock, travel agent
William Childs, optician
Ann R. Currie, teacher
Karl Ege, esq., attorney
David Fischel, astrophysicist
Steven H. Fleming, musician
J. Omar Flores, computer analyst
Robert C. Goyena, Sr., caterer
Don Hoover, carpenter
Dorothy B. Hunter, home economist
Spencer H. Klevenow, Jr., music teacher
Suella Klevenow, reading specialist
Jeanette Little, surgical assistant
John H. Little, Jr., wine maker
Carl L. Marburger, mathematician
Paul Monahan, fireman
John H. Pugh, encyclopedic mind
Chandler S. Robbins, ornithologist
David Rothfuss, thoroughbred horse trainer
Peter L. Schultz, linguist
Dick Schnacke, toy maker
John H. Steinway, piano maker
Dr. L. Donald Tamkin, orthodontist
Capt. Brian G. Traynor, police officer
Daniel J. Vitiello, esq., attorney
Karen H. Vitiello, thanatologist
Karen Yengich, journalist

5 21 32